Probabilities in the Galaxy

A Distribution Model for habitable Planets

Klaus Piontzik, Claude Bärtels

© Klaus Piontzik, Claude Bärtels, 2019
Probabilities in the Galaxy
A Distribution Model for habitable Planets

Production and Publishing:
Books on Demand GmbH, Norderstedt

ISBN 9-783-7528-5524-1

Klaus Piontzik

Klaus Piontzik (*1954) is an electrical engineer, mathematician and author. He can look back on a career of about 30 years as a project engineer in the industrial sector.
In the last 20 years he has increasingly specialized in electromagnetic fields, especially with regard to the Earth magnetic field.
In the last 10 years he also worked as an author (Lattice Structures of the Earth Magnetic Field, Planetary Systems I, Converting DNA in Colours and Tones, The Alien-Hypothesis, Odysseus 2013) and as a web author (www.pimath.de/eu).

Dr. Claude Bärtels

Dr. Claude Bärtels (*1956) is a biochemist and biophysicist. He can look back on a career of more than 25 years as a scientist, laboratory manager and division manager of international companies.
In recent years he has increasingly focused on the effects of electromagnetic fields and their effects on water and biological systems.
In the last 2 years he also co-authored (Planetary Systems I, The Alien-Hypothesis).

Probabilities in the Galaxy
A Distribution Model for habitable Planets

Tables of Content

Site

Introduction		8
1	**Planets in the Galaxy**	9
1.1	Detection of Planets	9
1.2	Data of the Kepler Satellite	10
1.3	Evaluation of Kepler Data	11
1.4	Sun-like Starsystems	11
1.5	G-Stars with Planets	12
1.6	G-Stars with habitable Planets	12
1.7	Probabilities for habitable Planets	14
1.8	Summary	14
2	**Evaluation of Catalogue Data**	15
2.1	Newer Catalogue Data for Exoplanets	15
2.2	Subearth	17
2.3	Superearth	18
2.4	Approximately Earth-great Planets	20
2.5	Approximately Earth-like Planets	21
2.6	Summary	23
2.7	Conventions and Notation	24
3	**„Earth 2.0"**	25
3.1	How many „Earth 2" are possible?	25
3.2	Case Distinctions	25
3.3	Consequences	27
3.4	„How many stars are there?"	28
3.5	Statistics	30
4	**Animated Planets in the Galaxy**	31
4.1	Planetary Assumptions for Life	31
4.2	„Earth 2" with Life	36
5	**Intelligent Species in the Galaxy**	38
5.1	Global Catastrophies	38
5.2	Planetary Dangers of Development	40
5.3	Intelligent Species on an „Earth 2"	44

		Site
6	**Civilizations in the Galaxy**	**46**
6.1	Development Levels of a Civilization	46
6.2	Distribution of Civilization Levels	56
6.3	Special Basic Model	60
6.4	Technological Civilizations	61
6.5	Comparable technological Civilizations	62
6.6	Space travelling Civilizations	64
6.7	Probabilities	65
7	**Survival of a Civilization**	**67**
7.1	Development Barriers of a Civilization	67
7.2	Age of a Civilization	68
7.3	Old Civilizations in the Galaxy	68
7.4	Temporal Distribution of Civilizations	70
7.5	Visitors	72
8	**General Basic Model**	**73**
8.1	Starsystems	73
8.2	Habitable Planets	73
8.3	„Earth 2"	75
8.4	Technological Civilizations	77
3.5	Other Civilizations	79
3.6	Comparison	80
8.7	Galactical habitable Zone	80
8.8	Probability Factors	82
9	**The Drake-Equation**	**83**
9.1	The classic Drake-Equation	83
9.2	Critics on the Drake-Equation	85
9.3	Carl Sagan	86
9.4	The modified Drake-Equation	87
9.5	Drake-Equation and General Basic Model	88
9.6	Corrected Values for the Earth	90
9.7	Corrections for Life and Intelligence	92
9.8	Corrections for the Basic Model	92
9.9	Other Civilizations	94
10	**The Seager-Equation**	**95**
10.1	The Equation from Sara-Seager	95
10.2	The extended Seager-Equation	96
10.3	The transformed Seager-Equation	98

		Site
11	**Equivalence of Considerations**	**100**
11.1	Equivalence	100
11.2	Corrected Values for the Earth	101
11.3	Corrections for Life and Intelligence	102
11.4	Corrections for the Basic Model	103
11.5	Other Civilizations	105
11.6	Result	106
12	**A General Approach**	**107**
12.1	Spectral Classes	107
12.2	Civilizations in the Galaxy	108
12.3	Technological Civilizations	110
12.4	Other Civilizations	111
12.5	Basic Model and General Approach	112
12.6	Drake-Equation and General Approach	113
12.7	Corrections for the Earth	114
13	**Lines of Evolution**	**115**
13.1	Lines of Development on the Earth	115
13.2	Convergent Development	117
13.3	Humanoids in Sun-like Systems	118
13.4	Corrected General Basic Model	119
13.5	General Approach	120
13.6	Working Hypothesis	121
13.7	Probabilities	122
14	**Approximately Earth-great planets**	**123**
14.1	Influence of Gravity	123
14.2	Life and Civilization	125
14.3	Non Sun-like Systems	128
14.4	Result	129
15	**Distributions**	**130**
15.1	Distribution of Raw Materials	130
15.2	Maximum Distribution of Civilizations	132
15.3	Distribution of Civilizations	133
15.4	Result	135
15.5	Building Set	136
	Table	**137**

APPENDIX

Site

16	**The SETI-Project**	**139**
16.1	The History of SETI	139
16.2	Signals	143
16.3	Operating Time of SETI	144
16.4	No Answer	144
16.5	Quantum Technology	145
16.6	Distribution of Starsystems	146
16.7	The best Case	147
16.8	Distances and Periods	148
16.9	Consequences	149
17	**The Fermi-Paradox**	**150**
17.1	The Considerations of Fermi	150
17.2	The Situation Today	150

Bibliography   **152**

Images Directory   **161**

List of Names   **165**
    Persons   165
    Telescopes, Radio Telescopes   166
    Institutions   166
    Satellites, Space Stations   166
    Astronomy   167
    Epochs   169
    Races   169
    Human Development   170
    Locations   171

Keyword Index   **172**

Introduction

The procedure used in the following chapters of the book is referred in mathematics as the axiomatic procedure. It forces us to formulate all the requirements clearly. This prevents hidden premises from creeping in.

Hypotheses are drawn from the analysis of existing empirical data of the Kepler telescope, which are called **approaches** and **axioms**. The statements derived from this are formulated as **theorems**. All axioms and theorems lead to an overall representation of the situation, which can then be used as a **working hypothesis**.

It should be noted here that the following paper is **not** an exact calculation method for determining the number of extraterrestrial civilizations, but **an estimation of size using statistical methods**, based on data that is as reliable or empirical as possible.
This paper also shows how to develop a systematic model for life, intelligence and civilization in the galaxy.

The theory presented in this book, i.e. all prerequisites, approaches, axioms and propositions mentioned in the chapters, are falsifiable in the sense of today's epistemology. Can therefore, in the future, be confirmed or refuted.
Thus the model formulated here (especially Equation System 6.3.3, Equation 8.4.2 and Equation 12.2.2) fulfils the property of falsifiability, in Popper's sense and thus represents a scientific hypothesis regarding the existence of alien civilizations in a galaxy.
Furthermore, the model is a fundamental contribution to exobiology, namely the existence of (intelligent) life in the galaxy.

For the sake of completeness and traceability, we have published almost the entire model. This makes it possible for everyone to recalculate all numbers themselves.
This stringing together of mathematics may be somewhat tedious for a normal reader, but it seemed necessary to us in order to obtain sufficient transparency and to make the derivations comprehensible.

1 – Planets in the Galaxy

1.1 - Detection of Planets

If one wanted to clarify how many civilizations there are in our galaxy, one would first look around for Earth-like planets in solar-like star systems - from a first human perspective. As will be shown later, life, intelligence and civilization could also have arisen on slightly different planets or star systems. Which will lead to an extension of the model. The first discovery of an exoplanet, in a solar-like star system, was made in 1995 using the radial velocity method. In the meantime, international studies are being conducted to detect exoplanets in other star systems within this galaxy. Various methods are used: [1]

Transit Method
If the orbit of a planet lies so that it passes exactly in front of the star from the view of the Earth, these coverings produce periodic subsidence in its brightness. They can be detected by brightness measurements of the star, while the exoplanet passes in front of its central star. These measurements are carried out using terrestrial telescopes such as the SuperWASP or, to a much greater extent, satellites such as COROT, Kepler or ASTERIA.

Radial Velocity Method
Star and planets move under the influence of gravity around a common center of gravity. The star moves on shorter distances than the planet because of its larger mass. If one does not look exactly perpendicular to this orbit from the Earth, this periodic movement of the star has a component in visual direction, which can be detected by observing the alternating blue and red shift of the star.

Astrometric Method
The movement of the star around the common center of gravity also has components across the viewing direction. These are detectable by exact measurement of its star locations relative to other stars. If the star mass and distance are known, the mass of the planet can also be specified, since the orbital inclination can be determined.

Run-Time of Light Method
This method is based on a strictly periodic signal from a central star or a central double star. Due to the influence of gravity, the center of gravity of a circulating planet shifts, resulting in a temporal shift in periodic signals.

Direct Observation

Since 2004 there have also been direct observations of exoplanets through the Hubble Space Telescope and the Very Large Telescope of ESO (European Organization for Astronomical Research in the Southern Hemisphere).

1.2 - Data of the Kepler Satellite

The search for exoplanets is currently being carried out by the Kepler satellite telescope. [2] Kepler is a NASA space telescope. [3] It was launched on 7 March 2009, to search for extrasolar planets. [1] [4] Erik Petigura from the University of California at Berkeley evaluated exactly 150,000 stars in 2013.

The Kepler telescope identified **42,000** of the **150,000** stars observed that resemble our sun, i.e. stars of spectral class G. Kepler has dis-

covered a total of **603** systems with planets in their orbits. **10** planets are about the size of the Earth and orbit their star in the so-called habitable zone, [5] where life-friendly temperatures prevail. [6]

To evaluate the Kepler data, fluctuations in the brightness of stars that indicate a **transit** of a planet in front of its sun are used.

Due to the different orbital inclinations of the planets against our line of sight, however, only a part of Earth-like planets are covered that can be observed from our direction. The geometric probability F_K of such a transit for the Earth is **0,47 %** ≈ 1/213 %. [2] [7]

So there could be a maximum of 213 times as many systems with planets. With 603 systems discovered, this results in 128,439 undiscovered planetary systems. That is 85.6 % of the observed stars. This means that a maximum of 85.6 % of all star systems in the galaxy could have planets.

1.3 - Evaluation of Kepler Data

150,000 star systems were examined. Statistically, this is a data population large enough to obtain significant numbers for the distribution of solar-like systems. However, only a small section of the sky was examined. In order to be able to transfer the data to the entire galaxy, a prerequisite must be met.

| 1.3.1 Axiom | The data of the Kepler satellite can be transmitted throughout the galaxy. |

The Kepler data of Erik Petigura are used in the following paper (chapters 1 to 7) to develop **distribution statistics for habitable planets, in solar-like systems, in the galaxy**.

1.4 - Sun-like Starsystems

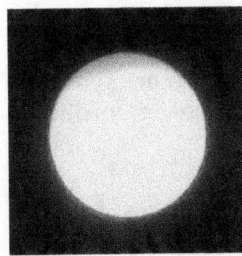

From **150,000** observed stars, **42,000** similar to our sun and belong to spectral class **G**. This corresponds to a share of 28 % of all observed stars. The probability factor extrapolated for the occurrence of G stars in our galaxy is therefore:

$$F_s = 0.28 = 42{,}000{:}150{,}000 = 7{:}25$$

Related to all star systems **A** in our galaxy, the number **N$_s$** of the solar-like star systems results with:

| 1.4.1 Equation | $N_s = A \cdot F_s$ |

with the following factors:

A = 100 - 300 Billion   Number of Suns in the Galaxy [8]
F_s = 0.28 = 7:25   Probability factor for solar-like stars

Insert the factors into Equation 1.4.1:

| 1.4.2 Theorem | The number of star systems, with Sun-like stars, in our galaxy is 28 to 84 billion. |

1.5 - G-Stars with Planets

From **42,000** Sun-like star systems, **603** were identified as planetary. This corresponds to a share of **1.435,7 %** with respect to solar-like systems. The probability factor is therefore:

$$F_p = 0.014{,}357 = 603{:}42{,}000 = \mathbf{201{:}14{,}000}$$

Related to all star systems **A** in our galaxy, the number N_p of solar-like star systems, with planets, results to:

1.5.1 Equation

$$N_p = N_s \cdot F_p$$
$$N_p = A \cdot F_s \cdot F_p$$

Applying all factors in Equation 1.5.1:

1.5.2 Theorem   The number of solar-like star systems, with planets, in our galaxy is probably 0.402 to 1.206 billion.

1.6 - G-Stars with habitable Planets

The **habitable zone** [5] is the **distance range** in which a planet must be located from its central star so that water can occur permanently in liquid form, as a prerequisite for life.

The exact position of a habitable zone in a solar system depends on the star type, i.e. on the radiation and the temperature of the star. The habitable zone can, simplified, be calculated from the luminosity of a star.

The **average radius** of this zone of any star can be calculated using the following equation:

$$d = \sqrt{\frac{L_{Star}}{L_{Sun}}}$$

d   is the average radius of the habitable zone in **AU**
L_{Star} is the bolometric luminosity of a star
L_{Sun} is the bolometric luminosity of the sun

In astronomy, bolometric brightness is a measure of the total luminosity of a celestial body, i.e. the luminosity integrated over the entire electromagnetic spectrum.
With a star twice as bright as the sun, the average radius is 1.4 AU.

    AU = Astronomical Unit = 149,597,870 Km = distance Sun-Earth

From **603** planetary systems observed by the Kepler telescope, planets in habitable zones could be detected in **10** systems. This corresponds to **1.658 %** of the investigated planetary systems. The probability factor is therefore:

    F_h = 0.016,58 = **10:603**

Related to all star systems **A** in our galaxy, the number N_h of solar-like star systems with habitable planets results to:

1.6.1 Equation
$$N_h = N_p \cdot F_h$$
$$N_h = A \cdot F_s \cdot F_p \cdot F_h$$

Insert the factors into Equation 1.6.1:

1.6.2 Theorem   The number of solar-like star systems, with habitable planets, in our galaxy is probably 6.666 to 20 million.

1.7 - Probabilities for habitable Planets

If you only look at the planets, then you can establish an equation for the probability of a habitable planet, in a Sun-like system.

1.7.1 Definition

$$F_{sph} = F_s \cdot F_p \cdot F_h$$

F_{sph} = 7:25 · 201:14,000 · 10:603
F_{sph} = 0.000,066,667 = **1:15,000**

Among **15,000** solar-like star systems there is one with at least one habitable planet.

1.8 - Summary

From the Kepler values, which were analysed by Erik Petigura, the following overall probabilities for **the distributions of planets in Sun-like systems in the galaxy** result:

Symbol	Rate	Factor	Designation
F_s	7:25	0.28	G-stars
F_p	201:14,000	0.014,357	G-stars with planets
F_h	10:603	0.016,583	G-stars with habitable planets
F_{sph}	1:15,000	0.000,066	habitable zone around a G-star

These factors are used as a **basis** for all further considerations and calculations.

2 – Evaluation of Catalogue Data

2.1 - Newer Catalogue Data for Exoplanets

In September 2015, a total of **1,952** exoplanets were known. According to the *„Habitable Exoplanets Catalog"* [1] there are **31** star systems with planets in habitable zones. **21** were superearth and only **10** were classified as approximately Earth-great. Only in **4** systems could you classify planets as approximately Earth-like. [2]
If 31 systems of 1,952 systems have habitable planets, this corresponds to a share of **1.588 %**, so: $F_{h2} = 0.015{,}88 = 31 : 1{,}952$.

NASA released on 9 May 2016, the latest data on the Kepler telescope. [3] [4] In the meantime, 1,284 new exoplanets have been discovered. Thus, a total of 2,325 exoplanets are now known.
There are probably rocky planets in **550** systems, like Earth. **9** planets are in a habitable zone.
Thus, the probability is $F_{h3} = 0.016{,}36 = 9 : 550$ for habitable planets.

According to Wikipedia, from 1 October 2017, a total of 3,671 exoplanets in 2,751 systems are known, including 616 systems with two to seven planets. [5]
According to Wikipedia, from 6 July 2018, a total of 3,801 exoplanets in 2,842 systems are known, including 633 systems with two to eight planets. [5]

According to the *„Habitable Exoplanets Catalog"* [1] of December 2017, **53** star systems with planets exist in habitable zones. **30** are superearths, **22** are classified as approximately Earth-great and a planet of the class subearth, thus planets which are smaller than the Earth. Only in **5** of **13** selected systems, planets could be classified as approximately Earth-like. [2]
If of 2,751 known star systems with planets, 53 systems with habitable planets exist, then the probability is $F_{h4} = 0.019{,}265 = 53 : 2{,}751$ for habitable planets.

The data from 2015 to 2017 show the minimum and maximum probability of a planet in the habitable zone:

2.1.1 Equation $0.015{,}88 \leq F_h \leq 0.019{,}265$

The reason for the differences is that we are dealing with a much smaller data population than in the Petigura investigation and the time intervals of the information are not constant, as well as the individual information sources are not synchronized. Small fluctuations are therefore unavoidable.

It should be noted once again that the exact position of a habitable zone in a solar system depends on the star type, i.e. on the radiation and the temperature of the star. The catalogue data for exoplanets did not always show the type of stars.

To make matters worse, the data material viewed here is not homogeneous in its statements. Some data are therefore difficult to compare, so a certain filtering had to be carried out.

A proven method is first of all to determine the limits, i.e. the maximum deviations, which is implemented with Equation 2.1.1.

The next step is to calculate the mean value from the three observation data and thus obtain:

$$F_{hm} = 0.017,572 \pm 0.001,693$$
$$F_{hm} \approx 1:57$$

Which corresponds quite well with the value F_{h1} = 0.016,588 = 10:603 from the Petigura investigation. The error is 4 %.

If you round the Petigura value a little from 10:603 to 10:600, you get 1:60, which comes quite close to the mean value of 1:57 from the observation data.

To simplify all further calculations, the rounded value from the Patigura analysis can be used, i.e.:

$$F_{h0} = 0.016,66... = 1:60$$

Overall, this results in a good agreement between the Petigura data and the observations made in the meantime and the resulting catalogue data.

The consistency of the data, as well as further information from the catalogue data can be used to derive further probabilities for subsets of habitable planets. These are the **subearths**, the **superearths**, **approximately Earth-great planets** and **approximately Earth-like planets**, which will be treated in the next sections.

2.2 - Subearth

Subearths are planets that have a smaller mass than Earth and are smaller in size. According to the *"Habitable Exoplanets Catalog"*, December 2017, there is **one** Subearth among **53** habitable planets. This corresponds to a share of **1.886 %**. The probability factor is thus F_{hsub} = 0.018,867 = 1:53.

2.2.1 Theorem   **1.886 % of all habitable planets, in solar-like star systems, are probably subearths.**

Then Equation 1.6.1 can be modified for solar-like star systems that have habitable planets and a relationship for subearths can be derived.

Related to all star systems **A** in our galaxy, the number N_{hsub} of the G-star systems, with habitable subearths, results to:

2.2.2 Equation

$$N_{hsub} = N_h \cdot F_{hsub}$$
$$N_{hsub} = A \cdot F_{sph} \cdot F_{hsub}$$

Starting point are 100 - 300 billion star systems in the galaxy, and inserting into Equation 2.2.2 provides:

2.2.3 Theorem   **There could be 125,786 to 377,358 habitable subearths in solar-like star systems in our galaxy.**

The probability of habitable subearths in solar-like star systems, in the galaxy, is then:

2.2.4 Definition $F_{sub} = F_{sph} \cdot F_{hsub}$

F_{sub} = 1:15,000 · 1:53
F_{sub} = 0.000,001.258 ≈ **1:795,000**

Among about **795,000** Sun-like star systems there is one with a habitable subearth.

In principle nothing speaks against the fact that life can also develop on subearths, at least in the form of flora.

It is doubtful, however, because of the low gravity and size only a thin atmosphere can form, whether on such planets really higher life, intelligence and civilization can develop.

2.3 - Superearth

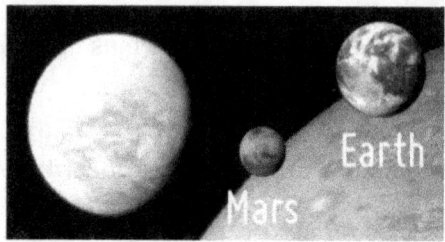

Superearths are planets with at least four times the Earth's mass. Most of the superearths are in a range of 4 to 20 Earth masses. However, there are also specimens of up to 40 Earth masses. Most of the superearths are about the size of the Earth or at most twice as large.

According to the „*Habitable Exoplanets Catalog*" of September 2015, 21 of 31 habitable planets are so-called superearths.
This corresponds to a share of 67.741 %. The probability factor is therefore F_{hsup} = 0.677,419 = 21:31.

According to the „*Habitable Exoplanets Catalog*" of December 2017, 30 of 53 habitable planets are superearths. This corresponds to a share of 56.603 %. The probability factor is therefore F_{hsup} = 0.566,037 = 30:53.

If you calculate the mean value from both information, you get:

$$F_{hsupm} = 0.621,723 \pm 0.055$$
$$F_{hsupm} = \mathbf{2,043:3,286}$$

2.3.1 Theorem 62.172 % of all habitable planets, in solar-like star systems, are probably superearth.

Then, from Equation 1.6.1 for solar-like star systems that have habitable planets, a relationship for superearth can also be derived.

Related to all star systems **A** in our galaxy, the number N_{hsup} of the G-star systems, with habitable superearths, results to:

2.3.2 Equation

$$N_{hsup} = N_h \cdot F_{hsup}$$
$$N_{hsup} = A \cdot F_{sph} \cdot F_{hsup}$$

Starting point are 100 - 300 billion star systems in the galaxy. Insert into Equation 2.3.2:

2.3.3 Theorem
There could be 4.144 to 12.434 million habitable superearths in solar-like star systems, in our galaxy.

The probability for habitable superearths in solar-like star systems, in the galaxy, is then:

2.3.4 Definition
$$F_{sup} = F_{sph} \cdot F_{hsup}$$

F_{sup} = 1:15,000 · 2,043:3,286
F_{sup} = 0.000,041,449 ≈ **1:24,126**

Among about **24,126** Sun-like star systems, one exists, with a superearth.

In principle, there is no reason why simple life, in the form of flora, can also develop on superearth.
However, because of the high gravity, it is doubtful whether intelligence and civilization can really develop on such planets. More likely are civilizations on approximately Earth-great planets and approximately Earth-like planets.

From the existing catalogue data, two further probabilities can be extracted for planets that can be used as life carriers, namely for **approximately Earth-great planets** and **approximately Earth-like planets**.

2.4 - Approximately Earth-great Planets

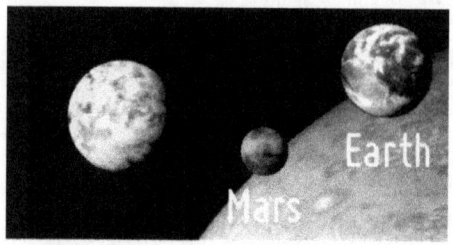

Approximately Earth-great planets are planets about the size of the Earth with at least one Earth mass. Almost all planets of approximately the size of the Earth found so far lie in a range of 1.3 to 4.8 Earth masses. Most of these planets are slightly larger than Earth, or at most twice as large.

According to the „Habitable Exoplanets Catalog" of September 2015, 10 of 31 habitable planets are about the size of the Earth. This corresponds to a share of 32.258 %. The probability factor is F_g = 0.322,58 = 10:31.

According to the „Habitable Exoplanets Catalog" of December 2017, 22 out of 53 habitable planets are about the size of the Earth. This corresponds to a share of 41.509 %. The probability factor is F_g = 0.415,094 = 22:53.

If you calculate the mean value from both information, you get:

$$F_{gm} = 0.368,837 \pm 0.046$$
$$F_{gm} = 1,212:3,286$$

2.4.1 Theorem   36.883 % of all habitable planets, in Sun-like star systems, are probably approximately Earth-great.

Then, from Equation 1.6.1 for solar-like star systems that have habitable planets, a relationship can also be extrapolated for approximately Earth-great planets.

Related to all star systems **A** in our galaxy, the number N_{hg} of solar-like star systems, with habitable, approximately Earth-great planets, results to:

2.4.2 Equation

$$N_{hg} = N_h \cdot F_g$$
$$N_{hg} = A \cdot F_{sph} \cdot F_g$$

Starting point are 100 - 300 billion star systems in the galaxy. Insert into Equation 2.4.2:

2.4.3 Theorem   There could be 2.458 to 7.376 million, approximately Earth-great, habitable planets in solar-like star systems, in the galaxy.

The probability for approximately Earth-great planets, in solar-like star systems, in the galaxy, is then:

2.4.4 Definition $F_{hg} = F_{sph} \cdot F_g$

$F_{hg} = F_{sph} \cdot F_g$
$F_{hg} = 1:15,000 \cdot 1,212:3,286$
$F_{hg} = 0.000,024,589 \approx \mathbf{1:40,668}$

Among about **40,668** Sun-like star systems there is one, with at least one habitable, approximately Earth-great planet.

2.5 - Approximately Earth-like Planets

Among the approximately Earth-great planets there exist still a number of planets which have a mass of Earth about the size of the Earth and which have a certain similarity to the Earth in the orbit time around their central star, as well as in the rotation time. Such planets are referred here as **approximately Earth-like planets**.

According to the „*Habitable Exoplanets Catalog*" of September 2015, 4 out of 10 habitable planets about the size of the Earth can be described as approximately Earth-like planets.
This corresponds to a share of 40 %. The probability factor is $F_a = 0.4 = 2:5$.

According to the „*Habitable Exoplanets Catalog*" of December 2017, 5 of 13 selected habitable approximately Earth-great planets can be described as approximately Earth-like planets. This corresponds to a share of 38.4 %. The probability factor is $F_a = 0.384,615 = 5:13$.

If you calculate the mean value from both information, you get:

$$F_{gm} = 0.392,307 \pm 0.007$$
$$F_{gm} = 51:130$$

2.5.1 Theorem   **39.23 % of all habitable planets, in Sun-like star systems, are probably approximately Earth-like.**

Then Equation 2.4.2 can be used to extrapolate a relationship for approximately Earth-like planets.
Related to all star systems **A** in our galaxy, the number N_{ha} of the solar-like star systems, with habitable, approximately Earth-like planets, results to:

2.5.2 Equation

$$N_{ha} = N_{hg} \cdot F_a$$
$$N_{ha} = A \cdot F_{sph} \cdot F_g \cdot F_a$$

Starting point are 100 - 300 billion star systems, in the galaxy, and inserting into Equation 2.5.2 provides:

2.5.3 Theorem   **There could be between 0.964 and 2.893 million approximately Earth-like habitable planets in solar-like star systems, in our galaxy.**

The probability for habitable, approximately Earth-like planets, in solar-like star systems, in the galaxy, is then:

2.5.4 Definition $F_{ha} = F_{hg} \cdot F_a$
                            $F_{ha} = F_{sph} \cdot F_g \cdot F_a$

F_{ha} $= F_{sph} \cdot F_g \cdot F_a$
F_{ha} $= 1:15,000 \cdot 1,212:3,286 \cdot 51:130$
F_{ha} $\approx \mathbf{1:103,664}$

As a result, only about every **103,664th** Sun-like star system produces a approximately Earth-like planet in the habitable zone.

2.6 - Summary

The planets found so far, usually about the same size as the Earth, can also be up to twice as large.
In the rotation time as well as the orbit time there are however considerable deviations and the mass of the found planets is always larger than that of the Earth. Gravity there is so high that no human being on Earth could live there any longer. Nevertheless, it cannot be ruled out that life in the form of flora and fauna could also develop there.

An „Earth 2" here means a planet whose size, rotation time and atmosphere are similar to Earth. Furthermore, it has oceans and continents and is so similar to the Earth that people could live there.

A real „Earth 2" has **not** yet been found.

All considerations presented so far are based on the evaluation of empirical data from the Kepler telescope and catalogue data. **They therefore represent the real situation**.

From the analysis of the catalogue data the following probabilities for the distributions of habitable planets result:

Symbol	Rate	Factor	Designation
F_{hsub}	1:53	0.018,867	Subearth
F_{hsup}	2,043:3,286	0.621,728	Superearth
F_g	1,212:3,286	0.368,837	approximately Earth-great planets
F_a	51:130	0.392,307	approximately Earth-like planets

2.7 - Conventions and Notation

In the following chapters planets are classified into these 6 categories:

2.7.1 Definition Planetary Categories

h	planets in (h)abitable zones
hsup	(h)abitable (Sup)erearth
hsub	(h)abitable (Sub)earth
g	approximatly Earth-(g)reat planets
	can be up to 2 times greater than the Earth
	can be up to 4 times larger than the Earth mass
a	(a)pproximatly Earth-like planets
	Rotation time, orbital time deviate from the Earth
	Mars belongs in this category
e	habitable, (e)arth-like planets („Earth 2")

The indices in all equations and considerations of this book are formatted, with the following meanings:

2.7.2 Definition Indices

s	(s)un-like
p	(p)lanet
h	(h)abitable Zone
sub	(Sub)earth
sup	(Sup)erearth
g	approximately Earth-(g)reat
a	(a)pproximately Earth-like
e	(e)arth-like
L	(L)ife
i	(i)ntelligent species
z	civili(z)ation
u	s(u)rvive civilization
m	hu(m)an, humanoid
x	Non Sun-like
RD	(R)ed (D)warfs

Convention: In the case of Sun-like systems, the index **s** may be omitted, since the Non-Sun-like systems are marked by an **x**.

3 – „Earth 2.0"

3.1 - How many „Earth 2" are possible?

Up to this point all considerations are based on the empirical data of the Kepler-telescope and catalogue data. Therefore they represent approximately the real situation. The probability F_e for a second Earth is still an uncertain parameter.

The term „**Earth 2**" derives from the phrase „Earth two point zero".

Developed from Equation 2.5.2 for solar-like star systems, with habitable, remote Earth-like planets, the number of the habitable „Earths 2" in our galaxy, according to this:

3.1.1 Equation

$$N_{he} = N_{ha} \cdot F_e$$
$$N_{he} = A \cdot F_{sph} \cdot F_g \cdot F_a \cdot F_e$$
$$N_{he} = A \cdot F_s \cdot F_p \cdot F_h \cdot F_g \cdot F_a \cdot F_e$$

3.2 - Case distinctions

A first estimate for the probability factor F_e can be developed.

CASE 1

The probability of encountering an approximately Earth-great planet in the habitable zone is $F_g = 0.368,837 = 1,212:3,286$.
The probability of finding an approximately Earth-like planet is $F_a = 0.392,3 = 51:130$. The probability for a second Earth could be in the same order of magnitude.
The probability of a second Earth is therefore set at a maximum of 30 %. The probability factor is therefore $F_{e1} = 0.3 = 3:10$. The probability of finding an „Earth 2" among habitable planets is therefore:

F_{gae1} = $F_g \cdot F_a \cdot F_{e1}$
F_{gae1} = $1,212:3,286 \cdot 51:130 \cdot 3:10$
F_{gae1} = $0.043,409 \approx 1:23$

That means: Among 23 habitable planets there should be an „Earth

2". So we should have found a second Earth by now. Since this has not yet happened (so far 53 habitable planets are known) the probability $F_{e1} = 0.3$ is too high and can therefore be omitted in the further considerations.

CASE 2

The probability of finding a approximately Earth-like planet in the habitable zone can be formulated as follows:

F_{e2} = $F_g \cdot F_a$
F_{e2} = 1,212:3,286 · 51:130
F_{e2} = 0.144,697 ≈ 1:7

The probability of a approximately Earth-like planet is 14.46 %. The probability for a second Earth is therefore chosen at a maximum of 10 %. The probability factor is $F_{e2} = 0.1 = 1{:}10$.

F_{gae2} = $F_g \cdot F_a \cdot F_{e2}$
F_{gae2} = 1,212:3,286 · 51:130 · 1:10
F_{gae2} = 0.014,469 ≈ 1:69

That means: Among 69 habitable planets an „Earth 2" should be found. Since 53 habitable planets are known, we should therefore be on the verge of finding a second Earth in the next few years.

CASE 3

The number of solar-like systems with planets determined by the Kepler telescope is 1.4357 %. The number of suns, with planets, in habitable zones is 1.658 %.
The probability of a second Earth is therefore chosen at a minimum of 1 %. The probability factor is $F_{e3} = 0.01 = 1{:}100$.

F_{gae3} = $F_g \cdot F_a \cdot F_{e3}$
F_{gae3} = 1,212:3,286 · 51:130 · 1:100
F_{gae3} = 0.001,446 ≈ 1:684

Among 684 habitable planets an „Earth 2" should be found.

3.3 - Consequences

Cases 2 and 3 of the case distinction can be used to obtain a minimum-maximum statement.

Inserting all values ($F_e = 0.1$) into Equation 3.1.1:

N_{he1} = (100 - 300)·10^9 · 1:15,000 · 1,212:3,286 · 51:130 · 1:10
N_{he1} = 96,465 – 289,395   habitable „Earth 2"

Inserting all values ($F_e = 0.01$) into Equation 3.1.1:

N_{he2} = (100 - 300)·10^9 · 1:15,000 · 1,212:3,286 · 51:130 · 1:100
N_{he2} = 9,646 – 28,939   habitable „Earth 2"

The two results can then be summarized to the following statement:

3.3.1 Theorem: There are probably 9,600 to 289,000 „Earths 2", in solar-like star systems, in our galaxy.

The probability for a habitable, Earth-like planet, in solar-like star systems, in the galaxy, is then:

3.3.2 Definition $F_{he} = F_{sph} \cdot F_g \cdot F_a \cdot F_e$
$F_{he} = F_s \cdot F_p \cdot F_h \cdot F_g \cdot F_a \cdot F_e$

F_{he} = $F_{sph} \cdot F_g \cdot F_a \cdot F_e$
F_{he} = 1:15,000 · 1,212:3,286 · 51:130 · (1:10-1:100)
F_{he} = 1:1,036,643 – 1:10,366,433

Only every **1.036 - 10.366 millionth** star system has a truly Earth-like planet.

If you only look at the habitable planets about the size of the Earth, then you can establish a probability for the similarity to an „Earth 2".

3.3.3 Definition $\boxed{F_{gae} = F_g \cdot F_a \cdot F_e}$

Two values can be generated:

F_{gae1} = 1,212:3,286 · 51:130 · 1:10 ≈ **1:69**

$F_{gae2} = 1{,}212{:}3{,}286 \cdot 51{:}130 \cdot 1{:}100 \approx \mathbf{1{:}691}$

According to Definition 1.7.1:

$$F_{sph} = F_s \cdot F_p \cdot F_h$$

Then Equation 3.1.1 for an „Earth 2" can also be expressed as follows:

3.3.4 Equation

$$N_{he} = A \cdot F_s \cdot F_p \cdot F_h \cdot F_g \cdot F_a \cdot F_e$$
$$N_{he} = A \cdot F_{sph} \cdot F_{gae}$$

$F_{sph} = 1{:}15{,}000$ Factor for a habitable Zone
$F_{gae} = 1{:}69$ to $1{:}691$ Factor for Earth Similarity

3.4 - „How many stars are there?"

The question that arises here is: How many stars do you have to examine in order to find an „Earth 2"?
Two cases can be distinguished, since the probability F_e, for Earth-like planets, can be at least 0.01 and at most 0.1.

CASE 1 ($F_e = 0.1$)

The probability of finding an „Earth 2" among habitable planets:

$F_{gae1} = F_g \cdot F_a \cdot F_{e1}$
$F_{gae1} = 1{,}212{:}3{,}286 \cdot 51{:}130 \cdot 1{:}10$
$F_{gae1} = 0.014{,}469 \approx \mathbf{1{:}69}$

That means: **among 69 habitable planets**, an „Earth 2" can be found.

To find these habitable planets, about 4,140 solar-like star systems with planets would have to be studied. That's about 289,800 Sun-like star systems. This would require a total of 1.035 million star systems to be investigated with the Kepler telescope. This is about **6.9** times the amount of stars examined so far.

Consequence 1

Among about 1.05 million star systems there is probably a system that contains an „Earth 2". This corresponds to **6.9** times the number of stars that have been examined by the Kepler telescope so far.

CASE 2 ($F_e = 0.01$)

The probability of finding an „Earth 2" under habitable planets is:

$F_{gae2} = F_g \cdot F_a \cdot F_{e2}$
$F_{gae2} = 1{,}212{:}3{,}286 \cdot 51{:}130 \cdot 1{:}100$
$F_{gae2} = 0.001{,}446{,}9 \approx \mathbf{1{:}691}$

This means: **among 691 habitable planets** an „Earth 2" can be found.
To find these habitable planets, you need to study 41,460 solar-like star systems that have planets. Thus, 2,902,200 Sun-like star systems and a total of 10.365 million star systems would have to be observed and analysed. That is **69** times the amount of stars examined so far.

Consequence 2

Among about **10.365 million** star systems there is probably a system that contains an „Earth 2". This is **69** times the number of stars that have been examined by the Kepler telescope so far.

Both cases combined result in the following statement:

3.4.1 Theorem   **The 6.9 to 69 times the number of stars that were examined with the Kepler telescope by 2013 are still needed to find an „Earth 2".**

At present, an almost exponential detection of planets is taking place through newer satellites and refined technologies. It is to be expected that a second Earth will be found within the **next decades**.

3.5 - Statistics

This results in the following probabilities for the distributions of solar-like star systems with habitable planets:

Symbol	Rate	Faktor	Bezeichnung
F_s	7:25	0.28	G-stars
F_p	201:14,000	0.014,357	G-stars with planets
F_h	10:603	0.016,583	planets in habitable zone
F_g	1,212:3,286	0.368,837	approximately Earth great planets
F_a	51:130	0.392,307	approximately Earth-like planets
F_e	1:100 – 1:10	0.01-0.1	Earth-like planets
F_{sph}	1:15,000	0.000,066	habitable zone around a G-star
F_{gae}	1:691 – 1:69	0.0014-0.014	Earth Similarity

Statistically, about **10** „Earth 2" were needed to obtain **significant** probabilities.
This would require the investigation of 69 to 691 times the number of stars that have been recorded by the Kepler satellite so far.
In 2013, 150,000 star systems were investigated with the Kepler telescope. Statistically, this is a data population large enough to obtain significant numbers. It can be assumed that with further investigations with larger populations and extended measuring methods, the determined probabilities (up to F_e) for Sun-like systems will change only slightly.
If the data is sufficiently significant, the probability factors are transformed into simple distribution or frequency values, making the probability model a simple distribution model.

Starting from the current speed at which exoplanets are recorded, however, it may take several years or decades before empirically significant figures are available.
The time until the discovery of a second Earth is even a measure of the distribution: the longer it takes to find an „Earth 2", the lower is the probability of F_e for a second Earth.

4 – Animated Planets in the Galaxy

4.1 - Planetary Assumptions for Life

In the previous chapter it could be clarified, how many Earth-like planets exist, which could offer a basis for life. The prerequisite for life is the presence of **basic building materials** such as carbon, nitrogen, oxygen, sulphur, phosphates and trace elements. And water, of course.

4.1.1 Axiom   If the basic building materials and suitable reaction environments are present in the universe, then the basic building blocks of life, such as amino acids, also arise there.

4.1.2 Axiom   The basic building blocks of life are created, under appropriate conditions, all in the universe.

Planetary Assumptions for Life are:

1) **Stable Orbit in a habitable Zone**

An orbit in the **habitable zone** is necessary to obtain temperature conditions suitable for life. Furthermore, the orbit must have a certain **time stability**, otherwise the climate and weather conditions would change too drastically. Even small changes in the orbit cause long-period climatic changes, which can result in recurrent ice ages.
The distance of the planet must be regulated in such a way that gravity and size of the planet are dimensioned in such a way that firstly a stable atmosphere can develop and se-

condly that the triple point of the water is stable. This is a fixed area between solid, liquid or gaseous, which results from distance, gravity (mass of the planets to each other), as well as centrifugal and centripetal forces. [1]

2) Stable Axis of Rotation

A stable **rotation axis** is necessary to maintain regulated and stable seasons. The inclination of this axis determines the seasons. [2]

The inclination of the axle should not be too large and the precession should not be too large to avoid greater climate and weather changes.
The duration of the precession should also not be too short. The **inclination** of the Earth's axis is **23.44°** from Earth's orbit and the **precession** is **25,800 years**.

Apart from stabilizing the axis of rotation, the **moon** also has an influence on the Earth and thus on climate and weather via the tides of the oceans.

The moon has a diameter of **3,476 km** and is **384,400 km** away from the Earth.
After a sidereal month (27.32 days) the moon takes the same position to the fixed stars again. Since he also turns around himself once, he always turns the same side to the Earth.

The following options are available for a **stable** axis of rotation:

a) one or more moons are present
b) a two-planetary system
c) as the moon of a much larger planet

3) Stable Magnetic Field, Electric Field (trigger signals, volcanism)

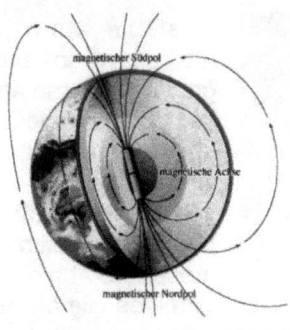

An **(electro)magnetic field** that has a certain temporal stability to a planet [3] is imperative to selectively protect against cosmic radiation and/or solar wind. Only defined parts of the particle flow of charged particles of solar origin (as well as parts of the light from IR, via visible to UV) are passed through. These are the "solar frequencies". Defined atmosphere windows exist for these specific frequencies. [4]

The Earth's magnetic field is generated by rotating magma masses inside the planet. This is one of the reasons why volcanism and plate tectonics occur simultaneously, which play an important role in shaping life. This magnetic field contains the "geomagnetic frequencies" (11.75 - 11.79 Hz). Another essential trigger signal is the "Schumann frequency" (7.83 Hz). The Schumann frequency is created by the formation of a standing wave with a cavity resonator frequency between the ionosphere and the Earth's surface. [5]

The frequency is a consequence of the distance ionosphere-Earth surface and circumference. In this respect, a stable ionosphere is even a prerequisite and the Schumann frequency is a consequence. These "geomagnetic frequencies" are essential signals in conjunction with the "Schumann" and "solar frequencies", so that a highly complex structure such as that of the hereditary substance as a "blueprint" and the structure of highly specific organs, enzymes, proteins and above all nerves can be clearly and perfectly performed and the function of all parts and elements can function together perfectly. [6]

These signals are necessary for the development of living beings to synchronize all structural, organic, nervous and mental processes. Overall, therefore, an electromagnetic field is required which has a certain temporal stability. As a result, the Geodynamo must remain active for billions of years. An example, if the Geodynamo comes to an early standstill, is Mars. Once like Earth with continents, oceans and atmosphere, it became a dried out planet due to the failure of the geodynamos and thus the magnetic field.

4) Stable Atmosphere
(light, shielding UV radiation, climate, weather)

A **stable atmosphere** is necessary to protect against UV radiation and smaller asteroids or comets. In addition, a better light distribution is achieved through the atmosphere. In addition, the atmosphere at night also determines climate and weather. [7] [8] A long-term stable atmosphere, with the associated climate systems, prevents an irreversible greenhouse effect, [9] as it has occurred on Venus. An atmosphere is also needed so that plants and living beings can develop.

5) Water (oceans, weather, climate)

Water is needed to create life and living beings need water to live. [10]
The presence of water requires oceans and these in turn influence the weather and climate of the planet. In total, **71 %** of the Earth's surface is covered by seas, i.e. the oceans and their secondary seas. About **3 %** of the water on Earth is fresh water. Most of it exists as frozen ice at the poles. Only about **0.03 %** of the world's water resources can be used as drinking water.

6) Continents (plants, climate, weather)

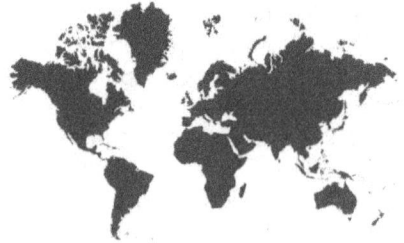

Volcanism and thus plate tectonic are generated by rotating magma masses inside the planet.
This creates **continents**. [11]

Continents are needed for plants and organisms to develop. The migration of continents changes fauna and flora. Especially at the edges of the plates volcanic activities can occur. Continents also have an impact on climate and weather.

7) **Basic Building Materials (chemical elements)**

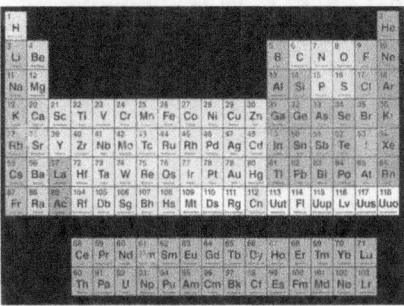

There are a number of chemical elements and their compounds, such as salts and minerals required to bring about life. [12]
There are **4,603** minerals on Earth.

8) **Basic Building Blocks of Life (amino acids)**

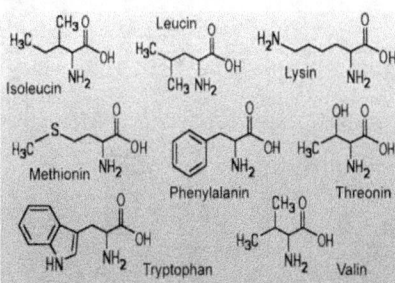

The building blocks of life are **amino acids**. These must be present for life to develop. [13]
There are **21** amino acids. Four of them are used in DNA, namely adenine, cytosine, guanine and thymine.

4.1.3 Axiom

If the planetary assumptions for life are given on a planet, then life develops there.

4.2 - „Earth 2" with Life

There are **8** components necessary to make life possible and develop, with all its possibilities. This can also cause 8 failures. Thus, the chance of life arises 1 to 9. That corresponds to a share of **11.11 %**. The probability factor is therefore $F_L = 0.111 = 1:9$.

Further developed, from Equation 3.1.1 results for the number N_{Le} animated "Earth 2" in Sun-like systems:

4.2.1 Equation

$$N_{Le} = N_{he} \cdot F_L$$
$$N_{Le} = A \cdot F_{sph} \cdot F_{gae} \cdot F_L$$
$$N_{Le} = A \cdot F_s \cdot F_p \cdot F_h \cdot F_g \cdot F_a \cdot F_e \cdot F_L$$

Since two values exist for F_e and F_{gae} a minimum-maximum statement can also be generated here:

Inserting all values ($F_e = 0.1$) into Equation 4.2.1:

$N_{Le1} = (100 - 300) \cdot 10^9 \cdot 1{:}15{,}000 \cdot 1{:}69 \cdot 1{:}9$
$N_{Le1} = 10{,}735 - 32{,}206$ habitable „Earth 2" with life

Inserting all values ($F_e = 0.01$) into Equation 4.2.1:

$N_{Le2} = (100 - 300) \cdot 10^9 \cdot 1{:}15{,}000 \cdot 1{:}691 \cdot 1{:}9$
$N_{Le2} = 1{,}072 - 3{,}216$ habitable „Earth 2" with life

The two results can then be summarized to the following statement:

4.2.2 Theorem   **The number of solar-like star systems, with a habitable „Earth 2" in the galaxy, which could carry life, is probably 1,072 to 32,206.**

The probability for an „Earth 2" with life, in a Sun-like star system, in our galaxy, is then:

4.2.3 Definition $F_{Le} = F_{sph} \cdot F_{gae} \cdot F_L$
$F_{Le} = F_s \cdot F_p \cdot F_h \cdot F_g \cdot F_a \cdot F_e \cdot F_L$

$F_{Le} = F_{sph} \cdot F_{gae} \cdot F_L$
$F_{Le} = 1{:}15{,}000 \cdot (1{:}69 - 1{:}691) \cdot 1{:}9$
$F_{Le} = 1{:}9{,}315{,}000 - 93{,}285{,}000$

Only every **9.315 - 93.285 millionth** star system then has an Earth-like, animated planet.

5 – Intelligent Species in the Galaxy

5.1 - Global Catastrophies

In order to estimate today what a species would have to be able to develop into a civilization, only the Earth is at our disposal. Therefore the development on this planet is considered in this chapter in order to be able to formulate an estimation of the probabilities.

1) About 2.4 billion years ago the "Great Oxygen Disaster" caused by cyan bacteria occurred. The appearance of free oxygen in the waters of the Earth and the Earth's atmosphere was poisonous for the anaerobic organisms of that time. Most anaerobic animals were extinguished. [1]
Today's, partly controversial, endosymbiont theory states that a phase took place in the prokaryotes (prokaryotes), which had survived the atmospheric change, could develop up to today's cell structure. This was achieved, according to the theory, in which they were incorporated into the developing multicellular (eukaryotes). The "How" is not described!
This cell formation is now known as the cell system known to us, as a cell with a nucleus with DNA and incorporated mitochondrium (single cell) with its own cell DNA, with which it lives in symbiosis and multiplies by cell division. This means that each of our body cells is a symbiont of "prehistoric development". So a single cell in the multi cell. Both "energy production systems" (ATP, NADP + NADPH) [Otto Warburg, 1931] are intertwined and have partly retained their ability to work autonomously. [2]

2) About 1 billion years ago was the "Snowball Earth", which is a hypothetical icing of the entire planet during the Hadean in the **Neoproterozoikum**. [3]
This resulted in the formation of eukaryotes, multicellular organisms, and sexual reproduction. [4]

3) About 80 % of all species of animals and plants died, including trilobites (trilobites), but also conodonts or brachiopods (caterpillars), about 485 million years ago at the end of **Cambrian** [5] Triggers were probably caused by climate change and / or sea level fluctuations. [6]

4) About 444 million years ago – in the upper **Ordovician** [7] –

there was a mass extinction which affected 50 % of all species. It disappeared, among other things, many Brachiopods. This extinction event is associated with the effects of the radiation of a near-earth supernova. [6]

5) About 360 million years ago, in the upper **Devon**, [8] another 50 % of all species died because the oxygen content in the water sank. It only survived animals that could adapt themselves or absorb oxygen outside the water. The time of the amphibians had begun. [6]

6) About 252 million years ago on the **Perm-Trias Border** [9] 95 % of all sea-dwelling species and about 66 % of all land-living species (reptile and amphibian species) died. A third of all insects also died out, the only known extinction of insects in the history of the Earth. The era of the Therapsids began. These are mammalian-like reptiles. Most scientists today make an extensive flood basin (Trapp), which is formed during volcanic eruptions, responsible for the "Siberian Trapp". [10]

7) About 200 million years ago, at the end of the **Triad**, [11] 50 to 80 % of all species died, among others. Almost all farm animals. It was followed by the age of dinosaurs.
Magmafreisations, before the break up of the Pangea continent, are suspected, or the poisoning of the shallow warm sea-waters by large quantities of hydrogen sulphide, after huge volcanic eruptions had released large quantities of carbon dioxide and sulfur dioxide. [6]

8) About 50 % of all animal species died, including dinosaurs, except for birds, some 66 million years ago on the **Cretaceous Tertiary Border** [12] (at the same time transition from the Earth's Middle Ages to the Earth New Age). It began the age of the mammals. The cause is the impact of a meteorite. [6]

9) About 33.9 million years ago, the so-called "Grande Coupure" cooled the global climate, which occurred at the turn of the **Eocene / Oligocene** [13] [14] (Priabonium / Rupelium border). [15] [16]
With related species dying and a change in the fauna, which fell victim to a large part of the primates (master animals), Palaeotheria (early horses), Creodonta (protruding animals)

and other animal groups. [17]

10) About 74,000 - 75,000 years ago, a **genetic bottleneck** developed in mankind. As a possible cause the Toba catastrophy theory is discussed by which humanity was reduced to a few thousand people when the Toba volcano erupted on Sumatra (Indonesia). [18]

11) About 13,000 years ago, since the end of the upper **Pleistocene**, [19] partly also in the **Holocene**, [20] the bulk of the Mega fauna of America, Eurasia and Australia died out in the course of a quaternary extinction.
There are indications of the impact of a meteorite, (or part of a comet), which reduced the large mammals about 13,000 years ago. [6]

These are together **11** events, in which the entire biosphere of the Earth could have been destroyed. In this respect, one can classify the causes of these events as **planetary development dangers**.

5.2 - Planetary Dangers of Development

If one takes these events or their causes as a basis and then asks oneself further which cosmic and planetary sources of danger come into consideration, then a whole list of possibilities results:

- **Cosmic Rays**
- **Gamma Flash**
- **Supernova**
- **Solar Eruption**
- **Asteroid Impact**
- **Comet Impact**
- **Ghost Planets, vagabonding Stars**
- **Ice Planet (Snowball Earth)**
- **Climate Change**
- **Atmospheric Change**
- **Change in Sea Level**
- **Volcanism**
- **Supervolcano**

1) Cosmic Rays

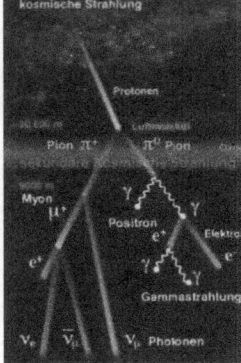

Cosmic Radiation is a high-energy particle radiation coming from the Sun, the Milky Way and distant galaxies.
It consists mainly of protons, electrons and fully ionised atoms. Approximately 1000 particles per square metre and second hit the Earth's outer atmosphere.
Particle showers are caused by alternating effects with the gas molecules. [21]

2) Gamma Ray Burst

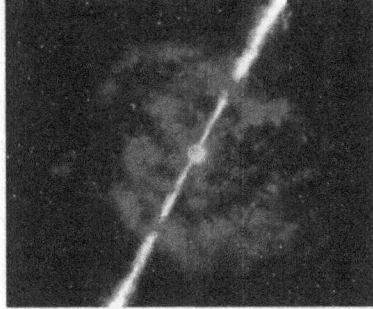

Gamma Ray Bursts are energy outbursts of very high power in the universe and emit large quantities of electromagnetic radiation. [22]

3) Supernova

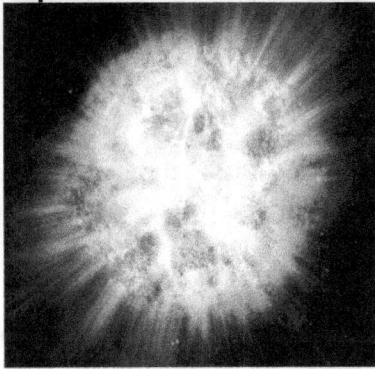

A **Supernova** is the brief and bright glow of a massive sun at the end of its lifetime.
Cause is the explosion of the sun. This will destroy the original star.
For a short time the luminosity of the star becomes as bright as a whole galaxy. [23]

4) Solar Flare

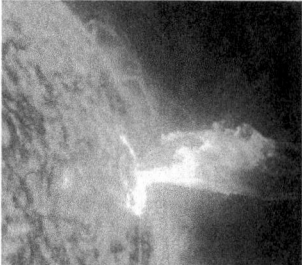

A **Solar Flare** is a structure of increased radiation within the chromosphere of the sun that is fed by magnetic field energy. [24]

5) Asteroid Impact

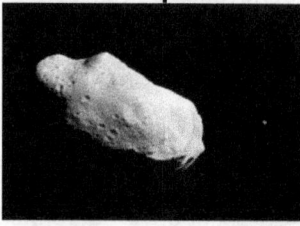

Asteroids are astronomic small bodies moving in so-called Kepler orbits around the sun. So far 742,836 asteroids are known in the solar system. [25]

6) Comet Impact

Comets are like asteroids the remains of the formation of the solar system.
They consist mainly of ice, dust and loose stones. They formed in the outer, cold areas of the solar system, in the so-called **Oort Cloud**. [26]

7) Ghost Planets, vagabonding Stars

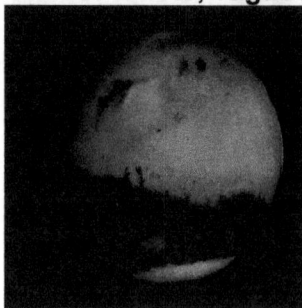

Ghost Planets are free planes in the galaxy, without connection to a star system. [27]
Vagabonding Stars are added, such as brown dwarfs, pulsars, neutron stars, magnetars and black holes. There are said to be several hundred free black holes in our galaxy.

8) Ice Planet (Snowball Earth)

The **Snowball Earth** means that during an ice age glaciers advance from the poles to the equator. The sea is then largely frozen over and the entire surface of the Earth is covered with ice.
This is said to have been the case 580 million years ago. [4]

9) Climate Change

Climate Change is the change in the Earth's climate, irrespective also of whether.
The causes are based on natural or also human (anthropogenic) activity.
A change in the climate is currently caused by the ongoing global warming, the greenhouse effect. Causes of the greenhouse effect include carbon dioxide, methane, ozone, chlorofluorocarbons, sulphur dioxide and nitrogen compounds. [28]

10) Change of Atmosphere

The **Atmosphere** is the gaseous envelope of the Earth. It has a high proportion of nitrogen, oxygen and a low proportion of argon. [29]

11) Change in Sea Level

Sea Level is the level at the surface of the sea. It corresponds approximately to an equipotential surface of the Earth's gravity field. [30]

43

12) Volcanism

Volcanism is understood to mean those geological processes and phenomena associated with volcanos. [31]

13) Supervolcano

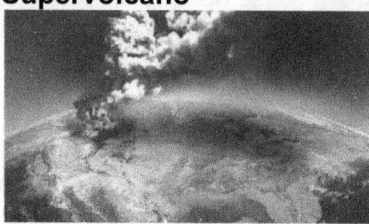

Supervolcanos, the size of their magma chamber, do not build up volcanic cones during eruptions, but leave huge calderas (craters) in the ground. [32]

These are a total of **13** possible planetary threats that have the potential to destroy the entire biosphere of the Earth.

5.3 - Intelligent Species on an „Earth 2"

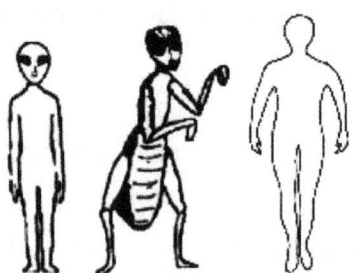

There are **13** causes listed that can destroy life on Earth globally. The chance that a species will survive and develop is therefore 1 to 14.

Only every **14th** planet that has life could produce a species capable of consciousness.
This corresponds to a share of **7.142 %**. The probability factor is then **F_i = 0.071,428 = 1:14**.

If one looks at the situation rather conservatively, then only planets come into question, which resemble an „Earth 2".

From the further development of Equation 4.2.1 for habitable Earths with life, the number of living "Earths 2" with an intelligent species is as follows:

5.3.1 Equation

$$N_{ie} = N_{Le} \cdot F_i$$
$$N_{ie} = A \cdot F_{sph} \cdot F_{gae} \cdot F_L \cdot F_i$$
$$N_{ie} = A \cdot F_s \cdot F_p \cdot F_h \cdot F_g \cdot F_a \cdot F_e \cdot F_L \cdot F_i$$

Since two values exist for F_e and F_{gae} a minimum-maximum statement can be generated here as well:

Inserting all values (F_e = 0.1) into Equation 5.3.1:

$N_{ie1} = (100 - 300) \cdot 10^9 \cdot 1{:}15{,}000 \cdot 1{:}69 \cdot 1{:}9 \cdot 1{:}14$
N_{ie1} = 767 – 2,300   „Earth 2" with intelligent Life

Inserting all values (F_e = 0.01) into Equation 5.3.1:

$N_{ie2} = (100 - 300) \cdot 10^9 \cdot 1{:}15{,}000 \cdot 1{:}691 \cdot 1{:}9 \cdot 1{:}14$
N_{ie2} = 77 – 230   „Earth 2" with intelligent Life

The two results can then be summarized to the following statement:

5.3.2 Theorem   The number of solar-like star systems, with an „Earth 2" in habitable zones, which have produced an intelligent species, is probably between 77 and 2,300.

The probability for a habitable „Earth 2", with intelligent life, in solar-like star systems, in our galaxy, then arises:

5.3.3 Definition $F_{ie} = F_{sph} \cdot F_{gae} \cdot F_L \cdot F_i$
$F_{ie} = F_s \cdot F_p \cdot F_h \cdot F_g \cdot F_a \cdot F_e \cdot F_L \cdot F_i$

F_{ie} = $F_{sph} \cdot F_{gae} \cdot F_L \cdot F_i$
F_{ie} = 1:15,000 · (1:69 – 1:691) · 1:9 · 1:14
F_{ie} = 1:130,410,000 – 1:1,305,990,000

Only every **130.41 millionth - 1.305 billionth** star system has an „Earth 2", with intelligent life.

6 – Civilizations in the Galaxy

6.1 - Development Levels of a Civilization

The age of the universe is given today at approximately 13.8 billion years. [1] About 600 million years later, our galaxy, the Milky Way, was born. Therefore, according to the current state of knowledge, the age of our galaxy is 13.21 billion years. [2] [3]
It took another 8.7 billion years for our solar system to develop. The age of the solar system and the Earth is therefore about 4.5 billion years. [4]
It took another 500 million years to develop life. The emergence of life on Earth is about 4 - 5 billion years back. [5] [6]
If "life" takes about half a billion years to develop on a planet, then the first civilizations may have appeared in our galaxy about 12.7 billion years ago. Since then, thousands of technological civilizations have emerged and may have passed. It is therefore probable that even today a high-tech civilization exists in the galaxy that has existed for several hundred thousand years. We would therefore still be young compared to these other, older civilizations.

  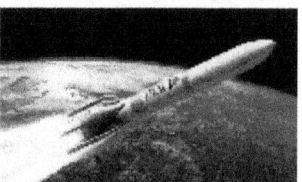

The development history of an intelligent species, from the origin of the species to high-tech civilization, can be divided into the following level of development or **civilization levels**:

0) **Prehistoric Level** (duration 20 - 25 million years)
   Splitting the old world apes into human and guenon relatives – about 23 million years ago
   Splitting of the apes and gibbons – 15 million years ago
   Separation of the chimpanzees from the hominini – 5.2 million years ago
   Australopithecus – 3.5 - 1.8 million years ago
   Upright walk

1) **Pre Level** (duration 2.5 - 3.5 million years)
   2.5 - 1.5 million years – Homo rudolfensis, Homo habilis
   2 - 1.5 million years – Homo erectus

Origin of the species, Stone Age
first stone tools about 2.5 million years ago

2) **Origin Level** (duration 300,000 - 350,000 years)
Neanderthals – 250000 - 30000 BC
340,000 years – oldest Homo sapiens find
Origin of the species, gatherers and hunters, Stone Age

3) **Primitive Level** (duration 30,000 - 40,000 years)
Stone Age, works of art, agriculture, settlements
First megalithic structures – Göbekli Tepe 9600 - 8000 BC

4) **Ancient Level** (duration 5,000 - 6,000 years)
Bronze Age, Iron Age, metals, cities, trade
Megalith culture 5000 - 2800 B.C.
Stonehenge 3100 - 1600 B.C.
Ancient races – Sumerians, Egyptians, Hittites, Cush, Minoans, Babylonians, Greeks, Romans, Chinese, Indians, etc.

5) **Intermediate Level** (duration 1,300 - 1,500 years)
Book printing, developments in science, art, medicine
End of Antiquity, Middle Ages, Early Renaissance

6) **Technological Level** (duration 350 - 450 years)
Renaissance, Baroque, Rococo, Age of Enlightenment, Modernity
Physics, Chemistry, Electrical Engineering, Aerospace

7) **Planetary Level** (1969 - Lunar Landing, 2009 - ISS Space Station, since 1970 - satellite missions in the solar system)
Postmodernity, colonization of the solar system

8) **Interstellar Level**
Interstellar space travelling in a galaxy

Other level of development are conceivable, such as:

9) **Galactic Level**
Space travelling between galaxies

10) **Cosmic Level**
Travelling in the Universe or Multiverse

However, it cannot be decided whether levels 9 and 10 are possible at all. It can also be assumed that such a level of development should lie far outside our present conceptual capacity and probably also occurs very rarely in the universe, so that only individual cases are to be expected here.

Since the first seven steps form a logical development chain and can be more or less empirically proven, these steps will serve as a basis for the next considerations.

An interesting conclusion can be drawn from the durations of human development levels, i.e. from level 2. To do this, you must first display the duration of a development level as a function of the development level:

Empirical	Level
300,000	2
30,000-40,000	3
5,000-6,000	4
1,200-1,500	5
300-400	6
Since 50 years	7

An approximation function can be determined with a table program (e.g. Excel). It therefore applies approximately for the **duration of a development stage**:

6.1.1 Equation

$$T_m = \frac{2 \cdot 10^7}{m^6} \quad m \geq 2$$

You can enter all values of the **development level function** in a table to get a better overview:

Empirical	Level	Function	rounded
300,000	2	312,500	310,000
30,000-40,000	3	27,434	28,000
5,000-6,000	4	4,882	5,000
1,200-1,500	5	1,280	1,300
300-400	6	428	430
Since 50 years	7	169	170
	8	76	80

The function is illustrated in the following two illustrations. See pages 50 and 51.

For level 7, the planetary level, according to the calculation, a period of 170 years is needed. Since humanity only about 50 years ago, by the moon landing, the seventh level has climbed, so we will still need about 120 years until we start to operate interstellar space travel.
So we won't have interstellar space travel until the 22nd century. With a development span of 80 years, i.e. to master interstellar space travel, the following predictions can be made for level 8:

6.1.2 Theorem   **The 22nd century could be the century of interstellar space travelling for mankind.**

If in the 22nd century mankind is engaged in interstellar space travel and can convince itself of the conditions.
As a result, all probability factors for the distribution of planets, life, intelligence and civilization that are being developed here will be determined with sufficient accuracy within the next two centuries.
Thus, the model developed here also transforms into a simple, empirical distribution model with regard to habitable planets, life, intelligence and civilizations in our galaxy.
In this respect, all derived parameters can or will be empirically verifiable in principle. Therefore the whole model is falsifiable and meets Popper's requirements. [29]

If Equation 6.1.1 is applied to the first stage of development, we obtain 20 million years. If one compares this with the evolution of man, the following facts result:
The oldest species of the genus Homo are Homo rudolfensis and Homo habilis. All known finds of Homo habilis were dated to an age of about 2.1 to 1.5 million years. [7] The finds of Homo rudolfensis surrendered to 1.9 million years. [8]
At 20 million years, the function for the development levels delivers a much higher value than the empirical data. Therefore, the original evel of the function is specified here with $m \geq 2$.
But one could also consider that the preliminary level of humanity took much longer than previously assumed.

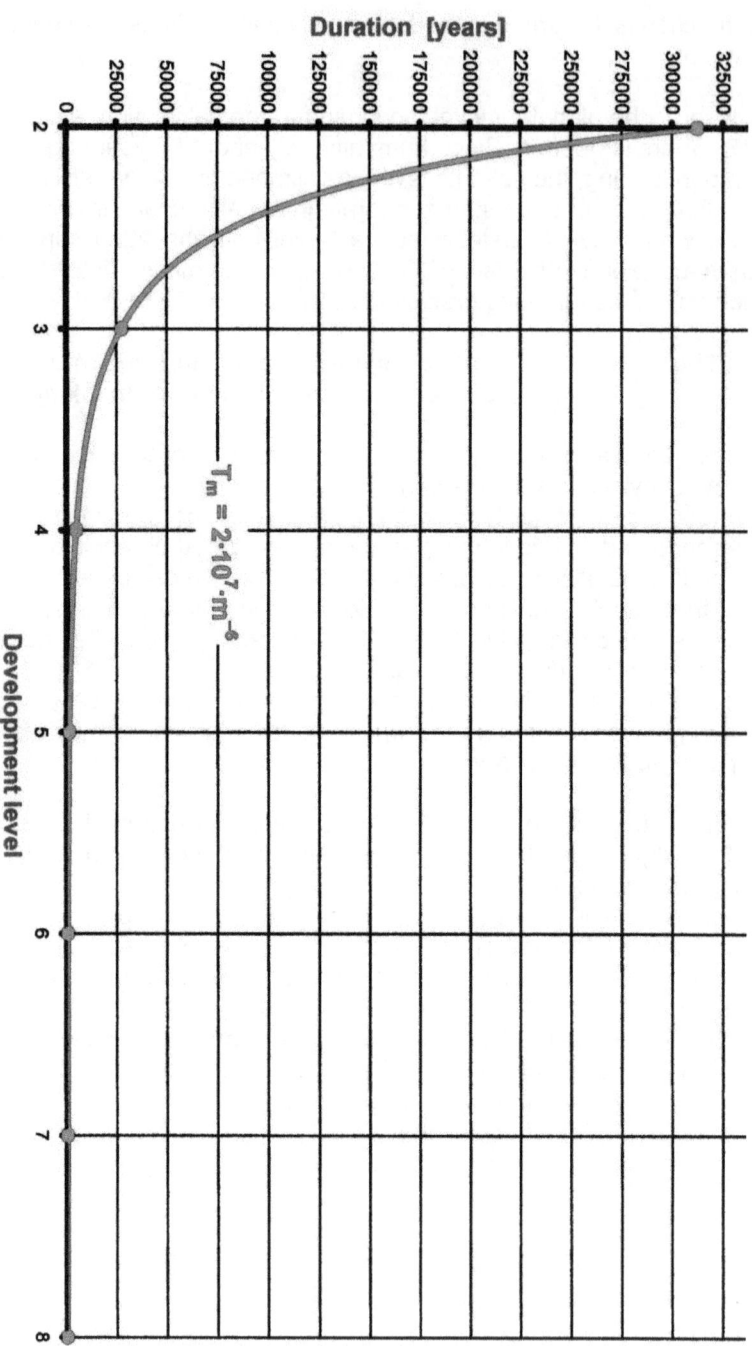

The relationship becomes more apparent when one logarithms the time axis.

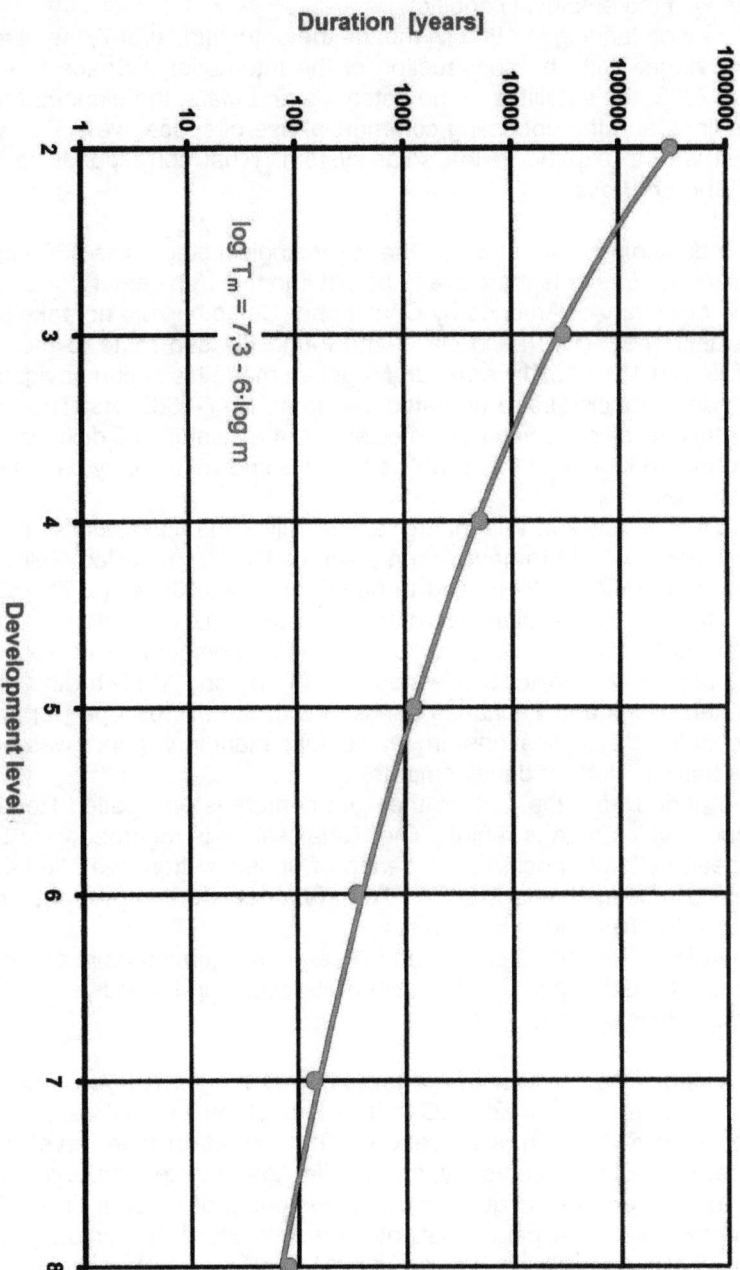

$\log T_m = 7{,}3 - 6 \cdot \log m$

With the given data, a review of the development levels can be made. One can compare the times given by the development level function with the empirical conditions.
The moon landing in 1969 [9] means the entry into **level 7**, the planetary stage. With the construction of the International Space Station ISS 2009, the satellites to the asteroids and Mars, the planned Mars landings and the upcoming commercial use of space, we are on the best way to expand in the solar system. What corresponds to the "**Planetary Level**".

The beginning of **level 6**, i.e. the technological stage, lies 430 years earlier at 1539 or is therefore to be put into the 16th century.
The discovery of America by Christopher Columbus did not take place until 1492. [10] Vasco Da Gama found the sea route to India in 1498. [11] 1519-1521 Fernando Magellan made his circumnavigation [12] and Francis Drake repeated this from 1577-1580. [13] The 16th century is also sometimes referred as the century of discoveries. Around 1600, half of the Earth's surface is known, but only part of the land area.
The thesis attack in Wittenberg, at which Martin Luther published his 95 theses, is said to have taken place in 1517. [14] Galileo Galilei lived from 1564-1642 [15] and through Johannes Kepler (1571-1630) the first scientific explanation of the solar system existed. [16]
The seafaring achievements of that time promoted technology in shipping. With Galileo's telescope 1608 [17] and 1595 by the lens makers Hans and Zacharias Janssen with the microscope [18], new technological applications arose. Kepler's planetary theory was also the basis for further developments.
The period from the 15th to the 16th century is also called Renaissance, which means rebirth. The Renaissance is regarded as a European cultural epoch, in the period of upheaval from the Middle Ages to modern times. [19] **The Renaissance thus represents the transition from level 5 to level 6**.
Therefore, the 16th century can be seen as a turning point at which humanity has begun the 6th level of its evolution towards a technological development.

The beginning of **level 5** is again about 1300 - 1500 years earlier than at stage 6, i.e. at 200 BC to 0, i.e. about the turn of time.
The turn of time thus represents the transition from level 4 to level 5. In European history, the Middle Ages marked the period between the end of antiquity and the beginning of modern times, i.e. approximately the period between the 6th and 15th centuries. [20] This fits in well with the "**Intermediate Level**".

The beginning of **level 4** is again about 5000 years earlier than at stage 6, i.e. at 5000 BC. This includes the Bronze Age (2200 to 800 BC) [21] as well as the Iron Age (from 1200 BC). [22] Here one also finds the ancient races, like Sumerians, Egyptians, Hittites, Cush, Minoans, Babylonians, Greeks, Romans, Carthaginians, Israelites, Chinese, Indians etc. [23] This fits in well with the **"Ancient Level"**.

The beginning of **level 3** is again about 28,000 years earlier than in stage 4, i.e. 33000 BC. The date of the beginning of agriculture is around 11000 B.C. The oldest settlement finds date from the same period. [24]
The first ceramic figures can be dated to at least 29,000 years. [25] About 35,000 years ago, cave painting was developed in southern France. The earliest finds of ivory carvings of figurines in Europe date from this period. The oldest evidence of a bone flute is also dated about 35,000 years. [26] This fits well with the **"Primitive Level"**.

The beginning of **level 2** is again about 310,000 years earlier than at stage 3, i.e. at 343000 BC. It's a good match with the oldest finds of Homo sapiens. [27]

This results in a good overall agreement between the development level function and the empirical data.
Which raises the question of how long it took humanity as a whole to develop. The sum of all development times T_M from **level 2** results:
Sum of all Development times T_m up to level 2:

$$T_M = \sum_{m=2}^{8} T_m = 342,980 \, years$$

Then T_M is the **average development time** or life span of a civilization.

Another way to determine the development time of a civilization is to calculate the area under the time function T_m, which corresponds to the total time of a civilization.
So you have to form the integral over the time function to get the total time span. The **total development time** of a civilization corresponds to the integral over the time function of a civilization.

6.1.3 Equation
$$T_{All} = \int_1^m T_m \, dm$$

$$T_{All} = \int_1^8 2 \cdot 10^7 \cdot m^{-6} \, dm = -\frac{2}{5} \cdot 10^7 \cdot m^{-5} \Big|_1^8$$

T_{All} = -2/5 · 10^7 · ($1^5 - 8^5$)
T_{All} = **3,999,877.93 years**

That is almost **4 million years** for the **total development time** of a civilization, from the preliminary level to interstellar space travel. This is quite in line with the empirical data.
If you take the time from level 2 to level 8, i.e. the time of purely human development, then the following applies:

T_{Homo} = -2/5 · 10^7 · ($2^{-5} - 8^{-5}$) = **124,877.92 years**

This is the minimum development time for an intelligent species from the primitive level to technological civilization.
Then it is to be expected that a civilization will exist 2 - 3 times longer, i.e. between 250,000 and 375,000 years. This could then be understood as the **average life span** of a civilization.
The sum of all development times T_m is ΣT_m = 342,980 years. Compared with the empirical 300,000 - 350,000 years that humans have needed, there is a good consensus here.
The time for prepress is the difference between the total time and the minimum development time. The preliminary level thus took **3.875 million years**, which is roughly in line with the empirical data.

The development stage function presented so far was determined by an approximation. This approximation is not the only solution function. Another possibility is presented here, which also includes level 1:

6.1.4 Equation $T_n = 10^{7{,}703 \cdot e^{-0{,}177 \cdot n}}$ n ≥ 1

Empirical	Level	Function	rounded
3,000,000	1	2,840,649	2,840,000
300,000	2	255,001	260,000
30,000-40,000	3	33,844	34,000
5,000-6,000	4	6,233	6,200
1,200-1,500	5	1,510	1,500
300-400	6	460	460
Since 50 years	7	170	170
	8	74	75

The total of all development times T_n:

$$T_N = \sum_{m=1}^{8} T_n = 3,137,295 \, years$$

This fits in well with the empirical data on the emergence of Art Homo. The oldest discovery dates back about 2.5 million years.

The sum of all development times T_n from level 2, i.e. the **average development time**, is added up:

$$T_N = \sum_{m=2}^{8} T_n = 297,292 \, years$$

This also agrees quite well with the empirical data on the development of Homo sapiens. The oldest find dates back 340,000 years.

However, the integral cannot be formed here. There is no default function for T_n.
T_n differ slightly from the previous time spans T_m. However, the deviations are so small that they have no influence on the previous considerations and results, as well as the resulting rates.

With the developmental model and the associated functions, an effective tool is now available to describe and classify the development of a civilization and to clarify the question of the **life span of a civilization**.

6.2 - Distribution of Civilization Levels

It can be assumed that the civilizations present in the galaxy will be distributed over the entire historical development levels. In a first approach one could assume that all civilizations are equally distributed over the levels of civilization.
On the other hand, it can be argued that the longer a civilization exists, the greater the probability of an all-destroying catastrophe. It is therefore more probable that the number of civilizations will decrease as the level of development increases. The following approach can be formulated from this:

6.2.1 Approach — The probability F_z for a civilization is inversely proportional to the level of development.

$F_z = 1:m$ and m = development level

This is illustrated once again in the following graphic.
If intelligent life has been created on a planet, it is 100 % likely that preliminary levels of civilizations have also been created, exactly those of level 1.
But already at level 2 the probability is only 0.5, at level 3 still 0.33, at level 4 only 0.25 etc.

It must now be demanded that the sum of all probabilities for the levels of civilization (m ≥ 2) is equal to one. This means that it is 100 % probable that at least one exists at the sum of all levels of civilization. So the rule is:

6.2.2 Equation $$S = \sum_{Z=2}^{8} F_Z = 1$$

However, this is not the case with the previous Approach 6.2.1, because totalling provides:

S = 1/2 + 1/3 + 1/4 + 1/5 + 1/6 + 1/7 + 1/8 = 481 : 280 > 1

However, probabilities cannot become greater than one. We must therefore take a tougher approach.

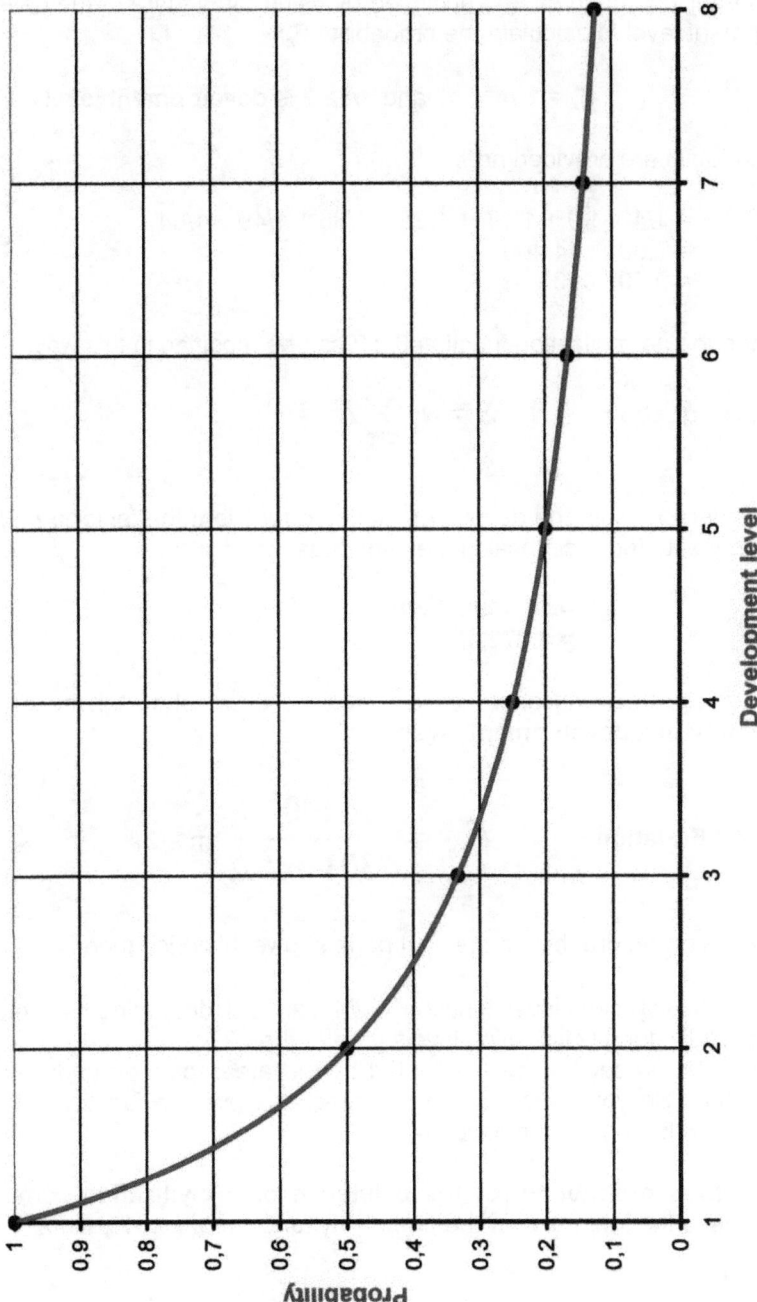

A better approach can be achieved by using the square of the development level to calculate the probability F_z:

$$F_z = 1 : m^2 \quad \text{and } m \geq 2 \text{ is development level}$$

The totals are provided here:

S = 1/4 + 1/9 + 1/16 + 1/25 + 1/36 + 1/49 + 1/64
S = 7,301 : 14,400
S = 0.507,013

Then the normalization function 6.2.2 can be modified in this way:

6.2.3 Equation $$S = a \cdot \sum_{Z=2}^{8} F_Z = 1$$

In order to adjust the developmental step function, the factor **a** must be equal to the reciprocal of the sum, thus:

a = 14,400 : 7,301
a = 1.972,332

Thus, the following approach can now be established for the **probability of a development level**:

6.2.4 Equation $$\boxed{F_{Zm} = \frac{14400}{7301 \cdot m^2}} \quad m \geq 2$$

The following graphic on the next page shows this once more.

The development level function 6.2.4 can still determine the total probability for all civilization levels greater than 2.
The total probability is represented by the area spanned under the function. So you have to form the integral above the function. This can then be formulated as follows:

The **total probability for the existence of a civilization** corresponds to the integral over the probability function of a civilization.

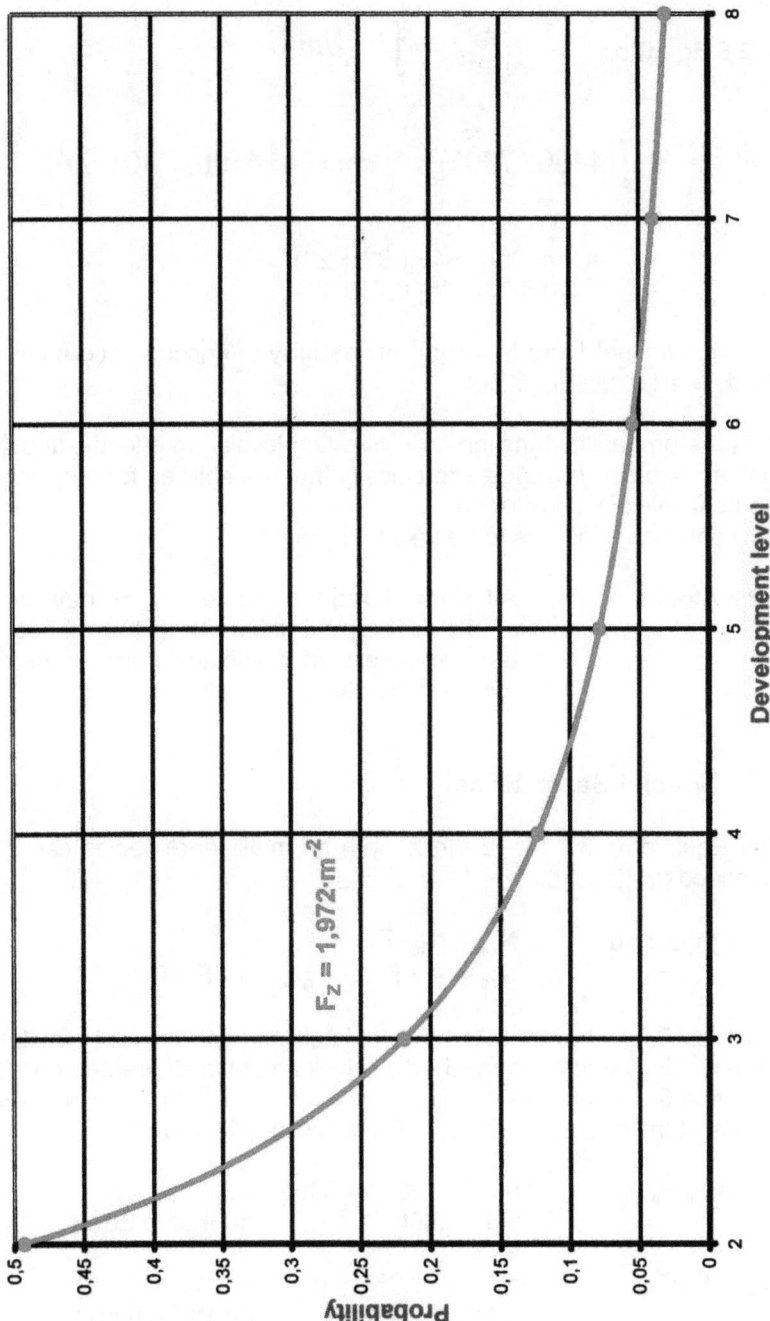

6.2.5 Equation
$$F_{All} = \int_{2}^{m} F_{zm} dm$$

$$F_{All} = \int_{2}^{8} 14400/7301 \cdot m^{-2} dm = -14400/7301 \cdot m^{-1} \Big|_{2}^{8}$$

F_{All} = -14400/7301 · (-8⁻¹ + 2⁻¹)
F_{All} = **5400/7301** = 0.739,624

This means that there is a **total probability** of encountering a civilization at all of about **74 %**.

With the probability function for civilization levels, an effective tool is now available to describe and classify the probabilities for the development levels of a civilization.
Now only one thing has to be taken for granted:

6.2.6 Axiom   All considerations on levels of development, development time, and the distribution of levels of civilization are transferable to extraterrestrial civilizations.

6.3 - Special Basic Model

The Equation 5.3.1 for „Earth 2" with an intelligent species can be extended to:

6.3.1 Equation $N_{ze} = N_{ie} \cdot F_z$
$N_{ze} = A \cdot F_{sph} \cdot F_{gae} \cdot F_L \cdot F_i \cdot F_z$

Equation 6.3.1 shows the number of intelligent species that may have technology and have reached a development level greater than 2. Equation 6.3.1 can be simplified a little further. To do this, you have to look at the probability factors. The following are known:

$F_{sph} = F_s \cdot F_p \cdot F_h$ = 7:25 · 201:14,000 · 10:603
                   = **1:15,000**   habitable Zone

$F_{gae} = F_g \cdot F_a \cdot F_e$ = 1.212:3.286 · 51:130 · (1:100 – 1:10)
                   = **1:702 bis 1:70**   Earth Similarity

If you only look at „Earth 2", you can still establish a probability for a civilization:

6.3.2 Definition $F_{Liz} = F_L \cdot F_i \cdot F_z$

F_z must be adapted to the quantities of civilization to be considered, which will become apparent in the following sections.

Equation 6.3.1 can then also be represented in this way:

6.3.3 Equation

$N_{ze} = A \cdot F_{sph} \cdot F_{gae} \cdot F_{Liz}$

$F_{sph} = F_s \cdot F_p \cdot F_h$ habitable Zone
$F_{gae} = F_g \cdot F_a \cdot F_e$ Earth Similarity
$F_{Liz} = F_L \cdot F_i \cdot F_z$ Civilization

The system of Equations 6.3.3 contains all planetary and biological factors that can influence the development of a civilization, by an intelligent species, on an „Earth 2", in a Sun-like system, in the galaxy.

Since we know one civilization, ours, Equation 6.3.3 can still be modified somewhat. It applies:

6.3.4 Equation $N_{ze} = A \cdot F_{sph} \cdot F_{gae} \cdot F_{Liz} \geq 1$

6.4 - Technological Civilizations

Only three of the eight levels of development represent higher **technological civilizations**, namely levels **6, 7, 8**.
This corresponds to a probability of:
$F_z = 14{,}400/7{,}301 \cdot (1{:}36 + 1{:}49 + 1{:}64)$
$F_z = 405{,}225/3{,}218{,}741 \approx 1{:}7.943$

$F_{Liz} = F_L \cdot F_i \cdot F_z$ $= 1{:}9 \cdot 1{:}14 \cdot 405{,}225/3{,}218{,}741$
     $= 0.000{,}999{,}171$
     $= 1 : 1{,}001$ technological civilization

Since two values exist for F_e and F_{gae} a minimum-maximum statement can also be generated here.

Inserting all values (F_e = 0.1) into Equation 6.3.3:

N_{ze1} = (100 - 300)·10^9 · 1:15,000 · 1:69 · 1:1,001
N_{ze1} = 97 – 290   „Earth 2" with a technological civilization

Inserting all values (F_e = 0.01) in the Equation 6.3.3:

N_{ze2} = (100 - 300)·10^9 · 1:15,000 · 1:691 · 1:1,001
N_{ze2} = 10 – 29   „Earth 2" with a technological civilization

The two results can then be summarized to the following statement:

6.4.1 Theorem   **There are probably 10 to 290 technological civilizations on an „Earth 2" in a solar-like star system, in our galaxy.**

The probability of a habitable „Earth 2", with a technological civilization, in solar-like star systems, in our galaxy, is then:

6.4.2 Definition $F_{ze} = F_{sph} \cdot F_{gae} \cdot F_{Liz}$

F_{ze} = $F_{sph} \cdot F_{gae} \cdot F_{Liz}$
F_{ze} = 1:15,000 · (1:69-1:691) · 1:1,001
F_{ze} = 1:1,036,035,000 - 1:10,375,365,000

Only every **1.036 - 10.375 billionth** star system produces an „Earth 2", with a technological civilization.

6.5 - Comparable technological Civilizations

Today's civilization has been at the 6th level for about 300 years, and we are standing with the moon landing in 1969, and the construction of the international space station ISS since 1998, at the beginning of the seventh stage.

Only two of the eight stages of development therefore represent comparable technological civilizations, namely levels **6** and **7**. This corresponds to a probability of:

F_z = 14,400:7,301 · (1:36+1:49) = **34,000:357,749 ≈ 1:10.522**

F_{Liz} = F_L · F_i · F_z = 1:9 · 1:14 · 34,000:357,749
                       = **1:1,326**   comparable civilizations

Inserting all values (F_e = 0.1) into Equation 6.3.3:

N_{ze1} = (100 - 300)·10^9 · 1:15,000 · 1:69 · 1:1,326
N_{ze1} = 73 – 219   comparable technological civilizations

Inserting all values (F_e = 0.01) into Equation 6.3.3:

N_{ze2} = (100 - 300)·10^9 · 1:15,000 · 1:691 · 1:1,326
N_{ze2} = 8 – 22   comparable technological civilizations

The two results can then be summarized to the following statement:

6.5.1 Theorem

There could be between 8 and 219 technological civilizations, on a „Earth 2", in a solar-like stellar system, in our galaxy comparable to ours.

Conclusion:
If the maximum comparable 219 civilizations are distributed evenly throughout the galaxy, thousands of Lightyears lie between them.
A signal from another civilization would therefore have taken a few thousand years and would also have had to be sent at the "right" time to reach us today. It would be an incredible stroke of luck to receive such a signal with the SETI Project.
Communication would also not be possible due to the long signal transmission time spans. This topic is dealt with in more detail in the appendix (The SETI Project).

The probability for a habitable „Earth 2", with a technologically comparable civilization, in solar-like star systems, in our galaxy, is then by Definition 6.2.5:

F_{ze} = F_{sph} · F_{gae} · F_{Liz}
F_{ze} = 1:15,000 · (1:69-1:691) · 1:1,326
F_{ze} = 1:1,372,410,000 – 1:13,743,990,000

Only every **1.372 - 13.743 billionth** star system has an „Earth 2", with a comparable civilization.

6.6 - Space travelling Civilizations

The prerequisite for a space travelling civilization is:

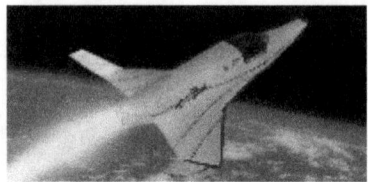

6.6.1 Axiom
There is a physics and resulting technology that allows interstellar journeys in short periods of time.

From a developmental point of view, only the highest level of civilization **8** has interstellar space travel. Therefore, the probability of finding such a species is:

F_z = 14,400:7,301 : 64 = **225:7,301** ≈ **1:32.448**

$F_{Liz} = F_L \cdot F_i \cdot F_z$ = 1:9 · 1:14 · 225:7,301
= **1:4,089**   space travelling civilizations

Inserting all values (F_e = 0.1) into Equation 6.3.3:

N_{ze1} = (100 - 300)·10^9 · 1:15,000 · 1:69 · 1:4,089
N_{ze1} = **24 – 71** space travelling civilizations

Inserting all values (F_e = 0.01) into Equation 6.3.3:

N_{ze2} = (100 - 300)·10^9 · 1:15,000 · 1:691 · 1: 4,089
N_{ze2} = **3 – 7**   space travelling civilizations

The two results can then be summarized to the following statement:

6.6.2 Theorem   There are probably 3 to 71 technological civilizations, on a „Earth 2", in a Sun-like star system, in our galaxy, which operate interstellar space travel.

The probability of a habitable „Earth 2" in solar-like star systems, in our galaxy, with a technological civilization operating interstellar space travel, is then by Definition 6.4.2:

F_{ze} = $F_{sph} \cdot F_{gae} \cdot F_{Liz}$
F_{ze} = 1:15,000 · (1:69-1:691) · 1:4,089
F_{ze} = 1:4,232,115,000 – 1:42,382,485,000

Only every **4.232 - 42.382 billionth** star system has an „Earth 2" with a space travelling civilization.

6.7 - Probabilities

If you look again at the factors F_{sph}, F_{gae} and F_{Liz} you can add the following:

F_{sph} = **1:15,000** is the probability for a planet in the habitable zone of a Sun-like star.

For F_{gae} = **1:691 – 1:69** only one range could be determined so far.

If up to 10 „Earths 2" will be found within the next 20 - 100 years, it will be possible to specify both parameters (F_{sph}, F_{gae}) with sufficient accuracy, i.e. statistically significant figures will then be available.

Values determined so far for F_{Liz}:

    F_{Liz} = **1:1,001** technological civilization
    F_{Liz} = **1:1,326** comparable civilization
    F_{Liz} = **1:4,089** space travelling civilization

The accuracy of these values will only be sufficiently clarified when mankind itself begins to operate interstellar space flight, i.e. according to Theorem 6.1.2 in 100 to 200 years, and can verify the distribution of life or intelligence and civilization in the galaxy itself.
As a result, all probability factors for the distribution of planets, life, intelligence and civilization will be sufficiently accurate within the next two centuries.
The system of Equations 6.3.3 contains all planetary and biological factors that can influence the development of a civilization, by an intelligent species, on an „Earth 2", in a solar-like star system, in the galaxy.
Subsequently, we will refer the system of equations for solar-like star systems derived up to this point as the „Special Basic Model".

If future further investigations lead to an ever better significance of the previous probability values, the probabilities of probability factors will change into simple distribution values.
Thus, the Basic Model is also transformed into a simple, empirical distribution model with regard to habitable planets, life, intelligence and civilizations in our galaxy.

Here once again all probabilities factors for life, intelligence and civilization.

Symbol	Rate	Factor	Designation
F_L	1:9	0.111...	planets with life
F_i	1:14	0.071,428	intelligent species
F_z	1:7.943	0.125,895	technological civilization
F_z	1:10.522	0.095,038	comparable civilization
F_z	1:32.448	0.030,817	space travelling civilization
F_{Liz}	1:1,001	0.000.999	technological civilization
F_{Liz}	1:1,326	0.000,754	comparable civilization
F_{Liz}	1:4,089	0.000,244	space travelling civilization

7 – Survival of a Civilization

7.1 - Development Barriers of a Civilization

In order for a civilization to develop so successfully that it reaches level 6 or higher, it must overcome at least four more difficulties. These are:

1) **Supply Shortage**
   In the history of mankind, some cultures such as the Maya or Anastasi have been destroyed by famine, persistent drought and/or climate change.
   Overexploitation [1], pollution [2] and overpopulation [3] are additional risk potentials.

2) **Epidemics, Pandemic**
   Historically, the plague and the "Spanish flu" were the great epidemics that have afflicted mankind so far. Today, for example, it is the Ebola virus on the African continent. Currently (Jan. 2016) an outbreak of the Zika virus (family of Flaviviridae) as a potential danger ("Zika Fever") is controversially discussed.
   All these pandemics have the potential to eradicate humanity. This is what happened to the Indians of North and South America when the Europeans appeared on their territory. [4]

3) **Wars**
   Many states and cultures have already fallen as a result of wars. Another potential risk of destruction over the past 60 years is the fact that the major industrial nations have sufficient destructive power to wipe out humanity several times. [5]

4) **Cataclysms**
   Regional disasters such as volcanism or lahar and pyroclastic flows or earthquakes caused by plate tectonics, as well as tsunamis, forest fires and hurricanes can develop into considerable difficulties in human development. [6]

5) **Internal Conflicts**
   Failure of the infrastructure due to hacker attacks or sabotage as well as social unrest and revolutions or civil war can lead to considerable difficulties in human development. [7]

Shortages of supplies, pandemics, wars, disasters and internal conflicts are partly caused by cultural factors and partly by environmental influences.
Every advanced civilization will have to overcome these obstacles in its development. There are five chances of failure. According to this, there is a chance for a civilization to survive from $F_u = 1:6$.

7.2 - Age of a Civilization

Today's humanity has existed for about 300,000 years [8] and it can be assumed that it will continue to exist for another 100,000 years. For comparison: the Neanderthal culture existed for about 250,000 years.
In view of the considerations in chapter 6.1 on the development stages, where the average service life with ΣT_m = 342,980 years and ΣT_n = 297,292 years, was determined in an initial approach, it can now be defined:

7.2.1 Axiom   The average life age of a technological civilization is set at L = 400,000 years (minimum).

You can assume that an old civilization is at least twice as old as that of of the average service life. This allows following definition:

7.2.2 Axiom   An <u>old</u> civilization is a culture that is at least one million years old.

7.3 - Old Civilizations in the Galaxy

The following equation results from the Basic Model 6.3.3:

7.3.1 Equation

$$N_{ue} = N_{ze} \cdot F_u$$
$$N_{ue} = A \cdot F_{sph} \cdot F_{gae} \cdot F_{Liz} \cdot F_u$$

Equation 7.3.1 contains all planetary, biological and civilizing factors that can influence the development of a civilization, by an intelligent species, on an „Earth 2", in a solar-like star system, in our galaxy.

It can be assumed that ancient civilizations emerge from the multitude of space travelling civilizations. Therefore $F_z = 225:7,301$ and $F_{Liz} = 1:4,089$.

Inserting all values ($F_e = 0.1$) into Equation 7.3.1:

$N_{ue1} = (100 - 300) \cdot 10^9 \cdot 1:15,000 \cdot 1:69 \cdot 1:4,089 \cdot 1:6$
$N_{ue1} = 4 - 12$ „Earth 2" with old civilizations

Inserting all values ($F_e = 0.01$) into Equation 7.3.1:

$N_{ue2} = (100 - 300) \cdot 10^9 \cdot 1:15,000 \cdot 1:691 \cdot 1:4,089 \cdot 1:6$
$N_{ue2} = 1 - 2$ „Earth 2" with old civilizations

The two results can then be summarized to the following statement:

7.3.2 Theorem   **There could be between 1 to 12 old technological civilizations, on an „Earth 2", in a solar-like star system, in our galaxy.**

The probability of a habitable „Earth 2" with an old technological civilization in solar-like star systems, in our galaxy, is then:

7.3.3 Definition $F_{ue} = F_{sph} \cdot F_{gae} \cdot F_{Liz} \cdot F_u$

$F_{ue} = F_{sph} \cdot F_{gae} \cdot F_{Liz} \cdot F_u$
$F_{ue} = 1:15,000 \cdot (1:69-1:691) \cdot 1:4,089 \cdot 1:6$
$F_{ue} = 1:25,392,690,000 - 1:254,294,910,000$

Only every **25.392 - 254.294** billionth star system has an „Earth 2" with an old technological civilization.

7.4 - Temporal Distribution of Civilizations

With the probabilities determined so far, it is possible to make a recalculation of how many civilizations could have existed since the formation of the galaxy. From this, a hypothetical **rate of civilization emergence** can be determined, which can provide information about the frequency of intelligent life in the galaxy.

R is the mean star formation rate per year in our galaxy. Depending on whether you are looking at galaxies, star clusters, or stellar nebulae, the value of **R** varies between 4 and 19. The mean is 11.5. [9]

T_g is the age of our galaxy, at 13.2 billion years. [10]

First, **S** is the number of stars that have formed since the galaxy was formed.

7.4.1 Equation $S = R \cdot T_g$

The next step is to determine the maximum number of civilizations that could have been possible since the galaxy was born. In doing so, the approximately Earth-like planets are taken as a measure.
F_{ia} is the probability of approximately Earth-like planets, with intelligent life, in this galaxy. According to the previous definitions and according to the rules of this model, the probability is:

$$F_{ia} = F_{sph} \cdot F_g \cdot F_a \cdot F_L \cdot F_i$$

Then the number **M** of intelligent species formed in the time span T_g can be determined as follows:

7.4.2 Equation
$$M = S \cdot F_{ia}$$
$$M = R \cdot T_g \cdot F_s \cdot F_p \cdot F_h \cdot F_g \cdot F_a \cdot F_L \cdot F_i$$

Equation 7.4.2 can still be converted to:

$$M = R \cdot T_g \cdot F_{sph} \cdot F_{ga} \cdot F_{Li}$$

The values determined so far are used for the factors:

R = 11.5   mean star formation rate
T_g = 13.2 billion years   age of our galaxy
F_s = 0.28   = 7:25   Sun-like star systems

F_p = 0.014,4 = 201:14,000 star systems with planets
F_h = 0.016,6 = 10:603 planets in habitable zones
F_g = 0.368,8 = 1,212:3,286 approximately Earth-great planets
F_a = 0.392,3 = 51:130 approximately Earth-like planets
F_L = 0.111... = 1:9 planets with life
F_i = 0.071,4 = 1:14 planets with intelligent species

F_{sph} = 0.28 = 1:15,000
F_{ga} = 0.142,8 = 1:7
F_{Li} = 0.007,9 = 1:126

The following result is obtained by inserting the values in Equation 7.4.2. Since the formation of the galaxy, **M = 11,474 intelligent species** have probably emerged. This allows the average rate of civilization development to be determined:

7.4.3 Equation
$$Z = \frac{T_G}{M}$$

$$Z = \frac{1}{R \cdot F_s \cdot F_p \cdot F_h \cdot F_g \cdot F_a \cdot F_L \cdot F_i}$$

This results in an **average rate of civilization development** of **Z = 1.15 million years per civilization**.

The rate of civilization formation is the mean cycle time between the appearance of two civilizations. This allows for a general definition of **old** civilizations.

7.4.4 Definition   A cycle of civilization is the period of time defined by the rate of civilization development.

7.4.5 Axiom   An <u>old</u> civilization is a culture that has survived at least one cycle of civilization.

7.4.6 Theorem   An <u>old</u> civilization is a culture that is at least 1.15 million years old.

7.5 - Visitors

According to Theorem 7.3.2, there are **1 - 12** old technological civilizations in our galaxy that operate interstellar space travel.
Really old civilizations, those that have existed for several million years, will have seen some species evolve, evolve and die out again over the course of their history.
Our civilization, with its little more than 300,000 years, is quite young in comparison with the "galactic" periods shown. We only reached technological level 6,200 years ago and are in the process of developing planetary level 7. On the interstellar level (level 8) we are at best newcomers.
Old alien civilizations may have visited and contacted humanity since the beginning of its development. It seems that they also left a number of traces (e.g. cave paintings) on humans. This makes the following hypothesis possible:

7.5.1 Theorem
Earth and humanity could, in the past, have been visited several times by alien species.

According to Theorem 6.6.1, there are probably **3 - 71** technological civilizations in our galaxy that operate interstellar space travel. It is therefore likely that they will still visit us today. The UFO phenomenon can be used here as an indication.
The Mutual UFO Network (MUFON) is an American organization dedicated to scientific research into the UFO phenomenon since 1969. It is one of the largest and oldest organisations worldwide in this field and has branches in several countries of the world, also in Germany, represented by MUFON-CES. [11]
UFO sightings have been documented since the 19th century and to date there are over **70,000** cases at MUFON alone in Hangar 1.
And part of it can be explained by the **extraterrestrial hypothesis**. This allows the following hypothesis to be made:

7.5.2 Theorem
There is a probability that our planet will still be visited by other species today.

8 – General Basic Model

8.1 - Starsystems

The „Special Basic Model" discussed in chapters 1 to 7 can also be applied to systems that are not similar to the sun.
Equation 1.4.1 for solar-like star systems can be generalized so that it applies to **any** number of stars of a spectral class in the galaxy.

Related to all star systems **A** in our galaxy, the number N_x of a set of star systems results with:

8.1.1 Equation $N_x = A \cdot F_x$

F_x is the probability of the occurrence of a set of stars that have a certain **property**.
A natural order of the stars is given by the system of **spectral classes**. [1] The spectral class of the G stars was already used in the Special Basic Model.
Equation 8.1.1 therefore applies to the star set of any spectral class. (see also chapter 12.1)
This means that the concept derived so far can be applied to all star sets for which observation data are available.

8.2 - Habitable Planets

If only the planets are considered, an equation for the probability for a **habitable planet** can be established in any star system. [2]

8.2.1 Definition $\boxed{F_{ph} = F_p \cdot F_h}$

So you can use the rounded values.

73

F_{ph} = 201:14,000 · 10:603 ≈ 1:70 · 1:60
F_{ph} = 1:4,200

Among **4,200** star systems there is probably one that has planets, at least one of which is a habitable planet.

Then applies to the set of all star systems of a spectral class that possess habitable planets:

8.2.2 Equation $N_{phx} = N_X \cdot F_{ph}$
$N_{phx} = A \cdot F_X \cdot F_{ph}$

It is important here that all factors can be determined in the long term by appropriate instruments of observation, i.e. empirically. According to Theorem 6.1.2, this should be the case in the next two centuries.

The set of all star systems N_{phxGal} in the galaxy, with planets and at least one in the habitable zone, is then the sum of all spectral classes:

$$N_{phxGal} = \Sigma N_{phx} = \Sigma (A \cdot F_X \cdot F_{ph})$$

Since **A** remains the same for all spectral classes, **A** can be taken from the sum:

8.2.3 Equation $\boxed{N_{phxGal} = \Sigma N_{phx} = A \cdot \Sigma (F_X \cdot F_{ph})}$

Strictly speaking, the probabilities F_p and F_h and thus F_{ph} would have to be determined individually for each spectral class.
However, since this is not yet possible, a rough calculation or maximum estimate can be achieved by assuming the same or similar frequencies of the planetary distributions in the spectral classes in a first approach.
If one assumes, in a first assumption, that other star systems, i.e. Non Sun-like stars, also have the same or similar planetary distribution, one can calculate how many systems could exist with habitable planets at most.

8.2.4 Approach In star systems that are Non Sun-like, there are probably the same or similar distributions, for habitable planets, as in Sun-like systems.

Provided that F_p and F_h remain the same for all spectral classes

(Approach 8.2.4), they can also be drawn from the sum (Equation 8.2.3):

$$N_{phxGal} = \Sigma N_{phx} \approx A \cdot F_p \cdot F_h \cdot \Sigma(F_X)$$

The sum of all spectral classes (for the probabilities F_X) results in the totality of the stars in the galaxy:

$$\Sigma(F_X) = 1$$

This results approximately from the Approach 8.2.4:

8.2.5 Equation $N_{phxGal} = \Sigma N_{phx} \approx A \cdot F_{ph}$

Starting point are 100 - 300 billion star systems, in the galaxy, and inserting all values into Equation 8.2.5 results:

8.2.6 Theorem   **The quantity of all star systems in the galaxy with planets and at least one in the habitable zone is probably a maximum of 23.81 to 71.428 million.**

8.3 - „Earth 2"

Analogous to the previous Special Basic Model we can generalize the Equations 3.1.1 and 3.3.4 for habitable „Earths 2" in Sun-like systems so that they apply to general star systems with **habitable, Earth-like planets**:

8.3.1 Equation $N_{hex} = N_{phx} \cdot F_{gae}$
$N_{hex} = A \cdot F_X \cdot F_{ph} \cdot F_{gae}$

The number of N_{hexGal} of all star systems with habitable, Earth-like planets in the galaxy is then the sum of all spectral classes:

8.3.2 Equation $\boxed{N_{hexGal} = \Sigma N_{hex} = A \cdot \Sigma(F_X \cdot F_{ph} \cdot F_{gae})}$

Assuming that the other star systems, i.e. Non Sun-like stars, also have the same or similar planetary distribution, it is possible to estimate how many systems with habitable „Earths 2" could exist.

8.3.3 Approach   In star systems that are Non Sun-like, there are probably the same or similar distributions, for Earth-like habitable planets, as in Sun-like systems.

The number N_{hexGal} of all star systems with habitable, Earth-like planets in the galaxy ($\Sigma F_x = 1$), is approximately equivalent to:

8.3.4 Equation $N_{hexGal} = \Sigma N_{hex} \approx A \cdot F_{ph} \cdot F_{gae}$

Inserting all values ($F_e = 0.1$) into Equation 8.3.4:

$N_{hexGal1} = (100 - 300) \cdot 10^9 \cdot 1{:}4{,}200 \cdot 1{:}69$
$N_{hexGal1} = 345{,}065 - 1{,}035{,}196$ habitable „Earth 2"

Inserting all values ($F_e = 0.01$) into Equation 8.3.4:

$N_{hexGal2} = (100 - 300) \cdot 10^9 \cdot 1{:}4{,}200 \cdot 1{:}691$
$N_{hexGal2} = 34{,}456 - 103{,}370$ habitable „Earth 2"

The two results can then be summarized to the following statement:

8.3.5 Theorem   **There could be between 34,000 and 1,035,000 „Earth 2" in our galaxy.**

8.3.6 Theorem   **There could be a maximum of up to 1 million „Earths 2" in our galaxy.**

The probability for habitable, Earth-like planets in any star system, in the galaxy, is then:

8.3.7 Definition $F_{hexGal} = F_{ph} \cdot F_{gae}$
$F_{hexGal} = F_p \cdot F_h \cdot F_g \cdot F_a \cdot F_e$

$F_{hexGal} = F_{ph} \cdot F_{gae}$
$F_{hexGal} = 1{:}4{,}200 \cdot (1{:}69-1{:}691)$
$F_{hexGal} = 1{:}289{,}800 - 1{:}2{,}902{,}200$

Only every **289,800 - 2,902,200th** star system produces an Earth-like planet, in the habitable zone.

It is left to the reader for practice to make the calculations for planets with life and intelligence himself. Otherwise, see the table at the end of the book (page 137).

8.4 - Technological Civilizations

Analogous to the previous Basic Model, Equation 6.3.3 for technological civilizations on an „Earth 2" in solar-like systems can be extended to such an extent that it applies to general star systems with habitable, Earth-like planets that have produced civilizations:

8.4.1 Equation $N_{zex} = A \cdot F_X \cdot F_{ph} \cdot F_{gae} \cdot F_{Liz}$

The number N_{zexGal} of all star systems, with habitable, Earth-like planets, with technological civilizations, in the galaxy is then the sum across all spectral classes:

8.4.2 Equation $N_{zexGal} = \Sigma N_{zex} = A \cdot \Sigma(F_X \cdot F_{ph} \cdot F_{gae} \cdot F_{Liz})$

Equation 8.4.2 is the extension of Equation System 6.3.3 and contains all planetary and biological factors that can influence the development of a civilization by an intelligent species, on an „Earth 2" in the galaxy.

In the further course of this paper, Equation 8.4.2 is therefore referred as the „General Basic Model".

Strictly speaking, the probabilities F_{ph}, F_{gae}, F_{Liz} would have to be determined individually for each spectral class. Since this is not yet possible today and in the near future, however, a rough calculation or maximum estimate can be achieved by assuming, in an initial approach, the same or similar distributions.

Assuming that Non Sun-like stars also have an equal or similar distribution for technological civilizations, one can make a rough estimate of how many systems with habitable „Earths 2" that carry civilizations could exist.

8.4.3 Approach In star systems that are Non Sun-like, there are probably the same or similar distributions, for Earth-like planets, with civilizations as in Sun-like systems.

The number N_{zexGal} of all star systems with habitable, Earth-like planets, with technological civilizations in the galaxy ($\Sigma F_x=1$), is then approximated:

8.4.4 Equation $\quad N_{zexGal} = \Sigma N_{zex} \approx A \cdot F_{ph} \cdot F_{gae} \cdot F_{Liz}$

Inserting all values ($F_e = 0.1$) into Equation 8.4.4:

$N_{zexGal1} = (100 - 300) \cdot 10^9 \cdot 1{:}4{,}200 \cdot 1{:}69 \cdot 1{:}1{,}001$
$N_{zexGal1}$ = **345 – 1,034** technological civilizations

Inserting all values ($F_e = 0.01$) into Equation 8.4.4:

$N_{zexGal2} = (100 - 300) \cdot 10^9 \cdot 1{:}4{,}200 \cdot 1{:}691 \cdot 1{:}1{,}001$
$N_{zexGal2}$ = **35 – 104** technological civilizations

The two results can then be summarized to the following statement:

8.4.5 Theorem The number of star systems in the galaxy, with Earth-like planets, in habitable zones that could carry civilizations, is most likely between 35 and 1034.

The probability of finding an Earth-like planet with a technological civilization is then:

8.4.6 Definition $\quad F_{zexGal} = F_{ph} \cdot F_{gae} \cdot F_{Liz}$

F_{zexGal} = $1{:}4{,}200 \cdot (1{:}69{-}1{:}691) \cdot 1{:}1{,}001$
F_{zexGal} = $1{:}290{,}089{,}280 - 1{:}2{,}905{,}102{,}200$

Only every **290.089 millionth - 2.905 billionth** star system then has an Earth-like planet, with a technological civilization.

8.5 - Other Civilizations

It is left to the reader to practice the calculations for comparable, space travelling civilizations. So here are the results:

8.5.1 Theorem   The number of star systems in the galaxy with Earth-like planets, in habitable zones that could carry comparable civilizations, is at most likely between 26 and 781.

8.5.2 Theorem   The number of star systems in the galaxy with Earth-like planets, in habitable zones that space travelling civilizations could carry, probably results at most between 9 and 253.

From the General Basic Model 8.4.2 the number N_{ueGal} of all star systems with habitable, Earth-like planets, with **old** civilizations in the galaxy, can be derived:

8.5.3 Equation $N_{ueGal} = A \cdot \Sigma(F_X \cdot F_{ph} \cdot F_{gae} \cdot F_{Liz} \cdot F_u)$

$N_{ueGal} \approx A \cdot F_{ph} \cdot F_{gae} \cdot F_{Liz} \cdot F_u$

Inserting all values ($F_e = 0.1$) into Equation 8.5.3:

$N_{ueGal1} = (100 - 300) \cdot 10^9 \cdot 1{:}4{,}200 \cdot 1{:}69 \cdot 1{:}4{,}089 \cdot 1{:}6$
$N_{ueGal1} = 14 - 43$ **old civilizations**

Inserting all values ($F_e = 0.01$) into Equation 8.5.3:

$N_{ueGal2} = (100 - 300) \cdot 10^9 \cdot 1{:}4{,}200 \cdot 1{:}691 \cdot 1{:}4{,}089 \cdot 1{:}6$
$N_{ueGal2} = 2 - 3$ **old civilizations**

The two results can then be summarized to the following statement:

8.5.4 Theorem   The number of star systems in the galaxy with Earth-like planets, in habitable zones that could support old civilizations, is at most likely between 2 and 43.

8.6 - Comparison

The Special Basic Model 6.4.1 provides **10 - 290** technological civilizations on habitable, Earth-like planets, in solar-like star systems.
The General Basic Model 8.4.5, provides **35 - 1,034** technological civilizations on habitable, Earth-like planets, in any star systems.

The Special Basic Model 6.5.1 provides **8 - 219** comparable civilisations on habitable, Earth-like planets, in solar-like star systems.
The General Basic Model 8.5.1, provides **26 - 781** comparable civilizations on habitable, Earth-like planets, in any star systems.

The Special Basic Model 6.6.2 provides **3 - 71** space travelling civilizations on habitable, Earth-like planets, in solar-like star systems.
The General Basic Model, 8.5.2 provides **9 - 253** space travelling civilizations on habitable, Earth-like planets, in any star systems.

The Special Basic Model 7.3.2 provides **1 - 12** old civilizations on habitable, Earth-like planets, in solar-like star systems.
The General Basic Model 8.5.4 provides **2 - 43** old civilizations on habitable, Earth-like planets, in any star systems.

The General Basic Model for stars of the galaxy provides more civilizations than the Special Basic Model for habitable „Earths 2" in solar-like systems. All in all, the result:

8.6.1 Equation

$$N_{CivGal} > N_{SCiv}$$

8.7 - Galactical habitable Zone

In 2001, the concept of a habitable zone in which life and life on Earth can develop was extended to galaxies. [3]

Originally, the concept of the galactic habitable zone referred to the chemical evolution of a galactic region, according to which there must be enough heavy elements in a region of a galaxy to create life. Most elements with larger atomic numbers than lithium are formed over time by nuclear fusion processes that take place inside the stars. When the stars die, these elements are released into interstellar space. This nucleosynthesis takes place faster in the inner regions of a galaxy than in the outer regions. This led to the assumption of a maximum radius of the galactic habitable zone. In the

course of time, the area should increase towards the outside.
These parameters are on the one hand very uncertain, on the other hand one can also question the premise to couple "life" with the star formation rate. So it may well be possible that the entire outer Milky Way is habitable. [4]

In the inner regions of a galaxy, the nucleosynthesis in which the heavy elements are produced is faster than in the outer regions.
Supernova explosions take place preferably in regions with active star formation, i.e. mainly in the spherical centre of a galaxy.
If a star with a planet is too close to a supernova explosion disturbs the atmosphere of the planet to such an extent that the planet is exposed to strong cosmic radiation that life could develop there permanently. [5]
For spiral galaxies, such as the Milky Way, the supernova rate rises to the inner region of a galaxy. Therefore, an inner radius of the galactic habitable zone can also be specified here.

The inner region of the galaxy accounts for about 8.5 % of the volume of the galaxy. Since the star density here is higher than in the outer regions, one can assume that about 10 - 15 % of all stars are here.
In this study, a maximum bandwidth of 100 - 300 billion stars is expected in the galaxy. This number would have to be corrected by the factor F_{GZ} = 0.8 - 0.9 to take the star density in the inner region into account.
With 100 - 300 billion stars, however, the factor 0.8 or 0.9 does not play a significant role. **It is already included in the error tolerance of the bandwidth of the number of stars** of this maximum observation.
Overall, the concept of a galactic habitable zone can therefore be neglected in this study.
If in the future (according to Theorem 6.1.2 in about 200 years) an empirical allocation of all data will be possible, the concept of the galactic habitable zone can also be included again.

8.8 - Probability Factors

Here are the values for all probability factors found:

Symbol	Rate	Factor	Designation
F_s	7:25	0.28	Sun-like stars
F_p	201:14,000	0.014,357	stars with planets
F_h	10:603	0.016,583	planets in habitable zones
F_g	1:1,212:3,286	0.368,837	approximately Earth-great planets
F_a	51:130	0.392,307	approximately Earth-like planets
F_e	1:100 – 1:10	0.01-0.1	Earth-like planets
F_{sph}	1:15,000	0.000,066	habitable planets, G-star
F_{ph}	1:4,200	0.000,238	habitable planets
F_{gae}	1:691 – 1:69	0.0014-0.014	Earth Similarity
F_L	1:9	0.111...	planets with life
F_i	1:14	0.071,428	intelligent species
F_z	1:7.943	0.125,895	technological civilizations
F_z	1:10.522	0.095,038	comparable civilizations
F_z	1:32.448	0.030,817	space travelling civilizations
F_u	1:6	0.166...	old civilizations
F_{Liz}	1:1,001	0.000,999	technological civilizations
F_{Liz}	1:1,326	0.000,754	comparable civilizations
F_{Liz}	1:4,089	0.000,244	space travelling civilizations

chapters 1 to 7 developed the Basic Model for Sun-like systems. In chapter 8 the Basic Model could be extended to all star systems of the galaxy, whereby the determined values are to be understood as maximum values.
Thus, a method is available that allows the number of technological civilizations in the galaxy to be estimated.
In the following chapters 9 and 10 two further possibilities of derivation for technological civilizations in the galaxy are discussed, namely the **Drake-Equation** and the **Seager-Equation**.

9 – The Drake-Equation

9.1 - The classic Drake-Equation

The **Drake-Equation** [1] is used to estimate the number of intelligent civilizations in our Milky Way. The equation was developed by **Frank Drake**, an US astrophysicist. [2]
In November 1960, for the first time, scientists from various disciplines met at Greenbank to discuss the probability of extraterrestrial intelligence and the search for it Frank Drake was responsible for the scientific content and conceivable topics.

For the conference Drake wrote some important points of discussion and wondered in what sequence the topics should be dealt with. All agenda items had the same importance, but they were not directly related. Drake assigned a symbolic factor to every meeting point and drew the individual factors into a simple multiplication formula to determine the number of highly developed and communicative civilizations, in the galaxy.

Frank Drake introduced this equation at the conference, and it is also referred as the *"Green Bank formula"* or the *"SETI equation"*. [3]
(Frank Drake uses other indices than defined in Definition 2.7.2)

9.1.1 Equation

$$N = R \cdot f_p \cdot n \cdot f_L \cdot f_i \cdot f_c \cdot L$$

R is the average star formation rate per year [4] in our galaxy. Depending on whether one is looking at galaxies, star clusters or stellar nebulae, the value for **R** varies between 4 and 19. The mean value is then **11.5**. The universal value is **1.45**.

f_p is the probability for a star system with planets. Here the value from the previous considerations is taken $f_p = F_p = 0.014,357 =$ **201:14,000**.

n is the number of planets in the habitable zone. Since probably only one planet in the habitable zone produces a civilization, **n** is set equal to one. The explanation is given below.

f_L is the probability for planets to have lives. Again, the value from the previous considerations (chapter 4.2) is taken, thus $f_L = F_L =$ **0.111... = 1:9**.

f_i is the probability for planets with technological civilizations.
The approach here is: $f_i = F_i \cdot F_z$.
The value from the previous considerations (chapter 5.3) for F_i is taken, thus $F_i = 0.071,428 = 1:14$.
The value from the previous considerations (chapter 6.4) for F_z is also taken here, thus $F_z = 0.125,895 = 1:7.943$.
Applying all values: $f_i = F_i \cdot F_z = 1:14 \cdot 1:7.943 = 1:111.203$.

f_c is the probability of the desire for communication. This value is set equal to **1**. The explanation follows later.

L is the life span of a communicable civilization. As defined in Axiom 7.2.1, the lifetime is set to a minimum of 400,000 years.

N is the number of extraterrestrial technological civilizations in the galaxy.

The Drake-Equation can now be partly expressed as a function of the parameters of the Basic Model:

9.1.2 Equation $N = R \cdot F_p \cdot n \cdot F_L \cdot F_{iz} \cdot f_c \cdot L$

Applying the values, in the Drake-Equation 9.1.2, provides:

$N = (1.45\text{-}19) \cdot 201:14,000 \cdot 1 \cdot 1:9 \cdot 1:111.203 \cdot 1 \cdot 400,000$
N = 9 - 109   extraterrestrial technological civilizations

Equivalent and thus comparable to the Drake-Equation, is Equation 8.4.5 from the General Basic Model, which describes all technological civilizations in the galaxy, on Earth-like planets. According to Theorem 8.4.5, the maximum number of star systems in the galaxy with Earth-like planets in habitable zones that could support technological civilizations is probably between **35 - 1,034**, and the Drake window is well located in the lower part of the General Basic Model window.

This results in a correspondence of the Drake window with the previous probability considerations from the Basic Model (chapters 1 to 7) or the General Basic Model (chapter 8).

It can be concluded that the value in the Basic Model for an Earth-like planet is closer to $F_e = 0.01$.

9.2 - Critics on the Drake-Equation

The main criticism of the Drake-Equation is ignited by the (previous) range of probability factors. Values can be used as desired for all factors. So that one can come to the conclusion that we are the only civilization in the galaxy or that we are just one civilization among many.
In the past, the Drake-Equation was therefore often referred as a collection of unknowns that remain unknown simply because they could not be determined. And therefore would leave far too much room for speculation and interpretation. [5] The Drake-Equation is therefore sometimes referred as a pseudo formula.
However, there is a misunderstanding here. The Drake-Equation is not a formula for calculating the number of civilizations in the galaxy. It is a **probability consideration** that serves to estimate **the order of magnitude** of the number of civilizations.

One must bear in mind here that most of the criticism comes from times when the claim that aliens existed was almost considered sacrilege or heresy.
However, as the data of the Kepler telescope show, the current information is already sufficient to approximately determine the factor f_p in the Drake-Equation.
According to Theorem 6.1.2, it can be assumed that in the next 200 years the remaining factors can also be sufficiently determined if humanity itself controls interstellar space travel.

Chapters 1 to 8 have so far shown that a differentiated approach is possible. In the next sections it will be shown that the Drake-Equation is compatible to the General Basic Model. The objections to the Drake-Equation can therefore only be seen as temporary challenges which will prove to be void in a (albeit distant) future.

9.3 - Carl Sagan

Carl Edward Sagan [6] (*1934 – †1996) was a well-known American astronomer and astrophysicist.

He also worked as a non-fiction author and writer. But it was only through his work as a television presenter that Carl Sagan gained a certain amount of publicity.

Sagan is considered a co-founder of **exobiology** and paved the way for the search for extraterrestrial intelligence **(SETI)**. [7]

Furthermore, Carl Sagan has contributed to many unmanned space missions that started in the second half of the 20th century and explored our solar system.

It was his idea to attach a message of humanity to a space probe that could also be understood by an alien intelligence. He realized this idea with the data disk **"Voyager Golden Record"** on the space probes **Voyager 1** and the almost identical **Voyager 2** in 1977. [8] [9]

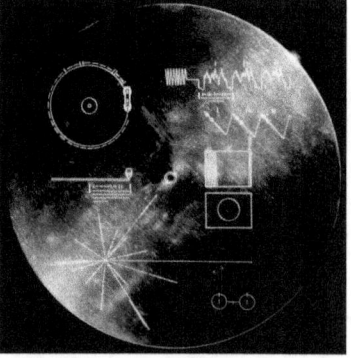

Carl Sagan also studied the Drake-Equation intensively and came to the conclusion that life outside the Earth is quite possible.

Due to the uncertainty of the probability factors at the time, he stated that the number of civilizations in the galaxy could be between **1** and **1 million**. [10]

9.4 - The modified Drake-Equation

However, there are still the following points of criticism about the Drake-Equation:

The factor **n** represents the number of habitable planets. In our solar system there are three habitable planets. And only one has life. The triple number of habitable planets does not mean that there are three times as many civilizations. Basically, you're just looking for a planet with a civilization. The factor **n** is therefore **1** and can therefore be omitted.

The factor f_c can also be set to **1**. If f_c = 0,5 this only means that half of civilizations do not want to communicate. But that doesn't make it **any less** civilizations. The factor f_c can also be neglected.

This allows to modify the Drake-Equation. The modified Drake-Equation is:

9.4.1 Equation $N = R \cdot f_p \cdot f_L \cdot f_i \cdot L$

And after transformation into the formula symbols of the Basic Model:

9.4.2 Equation $N = R \cdot F_p \cdot F_L \cdot F_{iz} \cdot L$

This is how the "**modified Drake-Equation**" can be formulated:

9.4.3 Equation $\boxed{N = R \cdot F_p \cdot F_{Liz} \cdot L}$

9.5 - Drake-Equation and General Basic Model

As chapter 9.4 shows, the Drake-Equation can be considered as equivalent to the General Basic Model 8.4.2 after fitting. For this to be exactly the case, both equations must be equivalent to each other.

8.8.4 General Basic Model: $\qquad N_{zexGal} = A \cdot F_{ph} \cdot F_{gae} \cdot F_{Liz}$

9.2.2 Transformed Drake-Equation: $\quad N \quad = R \cdot F_p \cdot F_{Liz} \cdot L$

It must then apply:

$N_{zexGal} \qquad\qquad = N$
$A \cdot F_{ph} \cdot F_{gae} \cdot F_{Liz} \quad = R \cdot F_p \cdot F_{Liz} \cdot L$
$A \cdot F_h \cdot F_{gae} \qquad = R \cdot L$

9.5.1 Equation $\qquad A \cdot F_h \cdot F_{gae} = R \cdot L$

Equation 9.5.1 can be used in several ways to check and correct the parameters. Here F_e offers the probability for Earth-like:

$$F_{gae} = R \cdot L / (A \cdot F_h)$$
$$F_e = R \cdot L / (A \cdot F_h \cdot F_{ga})$$
$$F_e = R \cdot L / (A \cdot F_h \cdot F_g \cdot F_a)$$

$F_e = 400{,}000 \cdot 11.5 / (100 - 300) \cdot 10^9 \cdot 10{:}603 \cdot 1{,}212{:}3{,}286 \cdot 51{:}130$
$F_e = 0.019{,}169 - 0.006{,}389$
$F_e = 1{:}52 - 1{:}156$

This results in the Earth Similarity F_{gae}:

$F_{gae} = 1{,}212{:}3{,}286 \cdot 51{:}130 \cdot (1{:}52 - 1{:}156)$
$F_{gae} = 1{:}359 - 1{:}1{,}078$

With Equation 9.5.1 the total number of stars in the galaxy can be determined:

$$A \cdot F_h \cdot F_{gae} = R \cdot L$$
$$A \qquad\qquad = R \cdot L / (F_h \cdot F_{gae})$$

$A = 400{,}000 \cdot 11.5 / 10{:}603 \cdot 1{:}359 = \mathbf{99.579 \cdot 10^9}$

$A = 400{,}000 \cdot 11.5 / 10{:}603 \cdot 1{:}1{,}078 = \mathbf{299.015 \cdot 10^9}$

This is in line with the given limits.

There are two ways to solve the equation. Solving the equation according to **L** or **R**.

Possibility 1

$L = A \cdot F_h \cdot F_{gae} / R$

$L = (100 - 300) \cdot 10^9 \cdot 10{:}603 \cdot 1{:}359 : 11.5 = \mathbf{401{,}689 - 1{,}205{,}068}$

$L = (100 - 300) \cdot 10^9 \cdot 10{:}603 \cdot 1{:}1.078 : 11.5 = \mathbf{133{,}772 - 401{,}316}$

On average, the life span of a civilization is:

$$L = 267{,}730 - 803{,}192 \text{ years}$$

This results in the average life span of a civilization:

$$L_m = 535{,}461 \text{ years}$$

This allows a correction for Axiom 7.2.1 to the lifetime of a civilization:

9.5.2 Axiom The average life span of a technological civilization is estimated at $L = 535{,}000 \pm 267{,}000$ years.

The maximum limits of a civilization's life span:

$$L = 133{,}000 - 1{,}205{,}000 \text{ years}$$

This is well in line with the 1.15 million year cycle of civilisation in chapter 7.4. The new maximum value allows a correction for Axiom 7.2.2 to the age of a civilization:

9.5.3 Axiom An <u>old</u> civilization is a culture that is at least 1.2 million years old.

Possibility 2

$R = A \cdot F_h \cdot F_{gae} / L$

$R = (100 - 300) \cdot 10^9 \cdot 10{:}603 \cdot 1{:}359 : 535{,}461 = \mathbf{8.627 - 25.881}$

$R = (100 - 300) \cdot 10^9 \cdot 10{:}603 \cdot 1{:}1.078 : 535{,}461 = \mathbf{2.873 - 8.619}$

On average, the star formation rate is:

$$R = 5.75 - 17.25$$

This results in the average star formation rate:

$$R = 11.5$$

This is exactly the same as the given mean value.

This observation allows to make a correction for the Drake-Equation. Insert the new values into Equation 9.1.2:

$N = (1.45\text{-}19) \cdot 201{:}14{,}000 \cdot 1{:}9 \cdot 1{:}111.203 \cdot 535{,}461$
$N = \mathbf{11 - 146}$ extraterrestrial technological civilizations

9.5.4 Theorem: **There are probably 11 to 146 technological civilizations in our galaxy.**

This result will be referred as **the corrected Drake-Equation**.

9.6 - Corrected Values for the Earth

From the last chapter 9.5 resulted a correction for the probability F_e of Earth-like, habitable planets:

9.6.1 Equation $1{:}156 \leq F_e \leq 1{:}52$

This results in the Earth Similarity F_{gae}:

9.6.2 Equation $1{:}1{,}078 \leq F_{gae} \leq 1{:}359$

This results in corrections for the number of „Earths 2", as well as for life, intelligence and civilization, as reported in the following sections.

Equation 3.1.1 for habitable „Earths 2" in solar-like systems:

$$N_{he} = A \cdot F_{sph} \cdot F_{gae}$$

Insert the new values into Equation 3.1.1:

$$N_{he} = (100 - 300) \cdot 10^9 \cdot 1{:}15{,}000 \cdot (1{:}359 - 1{:}1{,}078)$$

This results in the following correction for Theorem 3.3.1:

9.6.3 Theorem: There are probably **6,184 to 55,710** habitable „Earths 2", in solar-like star systems, in our galaxy.

In the smallest case, an „Earth 2" can be found among **359 habitable planets**. To find these habitable planets, you need to study 21,540 solar-like star systems that have planets. Thus, 1.507 million Sun-like star systems and a total of 5.385 million star systems would have to be observed and analysed. This is **36** times the number of stars examined so far.
In the largest case, an „Earth 2" can be found among **1,078 habitable planets**. To find these habitable planets, you need to study 64,680 solar-like star systems that own planets. Thus, 4.527 million Sun-like star systems and a total of 16.17 million star systems would have to be observed and analysed. That is **108** times the amount of stars examined so far.

All cases summarized results in the following statement:

9.6.4 Theorem The 36 to 108 times the number of stars that have been examined with the Kepler telescope by 2013 are still needed to find an „Earth 2".

After this (sobering) result it is to be expected that it will still take a **few decades** until one finds a second Earth.

9.7 - Corrections for Life and Intelligence

Equation 4.2.1 for habitable „Earths 2" with life in solar-like systems:

$$N_{Le} = A \cdot F_{sph} \cdot F_{gae} \cdot F_L$$

Insert the new values into Equation 4.2.1:

$N_{Le} = (100 - 300) \cdot 10^9 \cdot 1{:}15{,}000 \cdot (1{:}359 - 1{:}1{,}078) \cdot 1{:}9$

This results in the following correction for Theorem 4.2.2:

9.7.1 Theorem: There are probably 687 to 6,190 „Earths 2" with life, in Sun-like star systems, in our galaxy.

Equation 5.3.1 for habitable „Earths 2", with intelligent species in Sun-like systems:

$$N_{ie} = A \cdot F_{sph} \cdot F_{gae} \cdot F_L \cdot F_i$$

Insert the new values into Equation 5.3.1 applies:

$N_{ie} = (100 - 300) \cdot 10^9 \cdot 1{:}15{,}000 \cdot (1{:}359 - 1{:}1{,}078) \cdot 1{:}9 \cdot 1{:}14$

This results in the following correction for Theorem 5.3.2:

9.7.2 Theorem: There are probably 49 to 442 „Earths 2" with intelligent species, in solar-like star systems, in our galaxy.

9.8 - Corrections for the Basic Model

The **Special Basic Model** 6.3.3 for habitable „Earths 2", in solar-like systems, with a technological civilization is called:

$$N_{ze} = A \cdot F_{sph} \cdot F_{gae} \cdot F_{Liz}$$

Insert the new values into Equation 6.3.3:

$N_{ze} = (100 - 300) \cdot 10^9 \cdot 1{:}15{,}000 \cdot (1{:}359 - 1{:}1{,}078) \cdot 1{:}1{,}001$

This results in the following correction for Theorem 6.4.1:

9.8.1 Theorem There are probably 7 to 56 technological civilizations, on an „Earth 2", in solar-like star systems, in our galaxy.

The **General Basic Model** 8.4.2 for habitable „Earths 2" in the galaxy, with a technological civilization, is:

$$N_{zexGal} = \Sigma N_{zex} = A \cdot \Sigma(F_X \cdot F_{ph} \cdot F_{gae} \cdot F_{Liz})$$

$$N_{zexGal} \approx A \cdot F_{ph} \cdot F_{gae} \cdot F_{Liz}$$

Insert the new values into Equation 8.4.2:

$N_{zexGal} = (100 - 300) \cdot 10^9 \cdot 1:4,200 \cdot (1:359 - 1:1,078) \cdot 1:1,001$

This results in the following correction for Theorem 8.4.5:

9.8.2 Theorem   **The number of star systems in the galaxy, with Earth-like, habitable planets that could carry technological civilizations, is most likely between 22 and 199.**

Comparison Drake-Equation
Comparable here is the Drake-Equation 9.1.2 with **9 - 109** extraterrestrial technological civilizations. The Drake window coincides with the lower part of the window from the General Basic Model.
The corrected Drake-Equation 9.5.3 provides **11 - 146** alien technological civilizations. The corrected Drake window is identical to the lower part of the window from the General Basic Model.
Thus there is a good agreement between the **Drake-Equation** and the **General Basic Model**.

9.8.3 Theorem   **The General Basic Model and the Drake-Equation represent two mutually <u>equivalent</u> approaches.**

Since the number of star systems from the Basic Model is considerably larger than in the Drake-Equation, it can be concluded that the upper limit F_e is set too high for the probability of a second Earth.

9.9 - Other Civilizations

It is left to the reader to practice the calculations for comparable, space travelling civilizations. See also the table on page 137, therefore only the results:

9.9.1 Theorem   The number of star systems in the galaxy, with habitable, Earth-like planets that could support comparable civilizations, is at most likely between 17 and 150.

The General Basic Model 8.5.1 provides **26 - 781** comparable civilizations.

9.9.2 Theorem   The number of star systems in the galaxy, with habitable Earth-like planets that could support space travelling civilizations, is at most likely between 6 and 49.

The General Basic Model 8.5.2 provides **9 - 253** space travelling civilizations.

9.9.3 Theorem   The number of star systems in the galaxy, with Earth-like planets, in habitable zones that could support old civilizations, is at most likely between 1 and 8.

The General Basic Model 8.5.4 provides **2 - 43** old civilizations.

The corrected values entered in the General Basic Model (as seen) are here referred as the **Drake-corrected General Basic Model**.
All corrected Basic Model windows are identical to the lower parts of the windows from the General Basic Model.
This ensures good agreement between the **General Basic Model** and the **Drake-corrected General Basic Model**.

Again, the number of star systems from the Basic Model is considerably larger than in the Drake-Equation, so it may be concluded that the upper limit F_e is set too high for the probability of a second Earth.

10 – The Seager-Equation

10.1 - The Equation from Sara-Seager

Sara Seager [1] is a Canadian-American astrophysicist (*1971). She introduced a modified approach to the Drake-Equation. This approach is called the **Seager-Equation** and sometimes it is called the **Drake-Seager-Equation**.
In contrast to the Drake-Equation, their approach does not work with the star formation rate, but with a fixed set of stars, namely systems from the spectral class **M**.
Seager's approach is limited to the so-called M-stars, also called Red Dwarfs, and the future James Webb Space Telescope (JWST), [2] and the running TESS space satellite (Transiting Exoplanet Survey Satellite). [3]

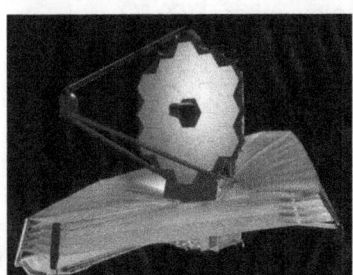

JWST   **TESS**

The James Webb Space Telescope is virtually the successor to the Hubble Telescope and is scheduled to be launched in 2021. The TESS satellite is gone on the journey on 18 April 2018 and is looking for exoplanets using the transit method. (Sarah Seager uses other indices than defined in Definition 2.7.2)

The Seager-Equation is:

10.1.1 Equation $\boxed{N = N^* \cdot f_Q \cdot f_{HZ} \cdot f_O \cdot f_L \cdot f_S}$

The following values for the probability factors are taken from a document that Sara Seager has placed on the Internet. [4]

N* stands for the number of M-stars (Red Dwarfs), which can be examined with the coming telescope JWST. (30,000 - 50,000)

f_Q stands for the proportion of quiet M-stars. The amount of stars, which repeatedly throw large amounts of gamma rays into space is 20 %.

f_{HZ} is the proportion of those systems that have a planet in the habitable zone. (circa 15 %)

f_O quantifies the number of planets that are visible for the JWST visibly past their star. (1 % of the potentially observable planets pass before their star, 10 % of which are close enough to the Earth for observation) (0.01 x 0.1= 0.001)

f_L represents the share of planets with life. The factor is set to 1 here because one assumes that life could be created on every habitable planet.

f_S is a measurable bio signature, in the atmosphere. (50 %)

Substituting the values in the Equation 10.1.1 yields:

N = (30,000 bis 50,000) · 0.8 · 0.15 · 0.001 · 1 · 0.5
N = 1.8 – 3   technological civilizations

According to Sara Seager, **N = 2**. This result shows that intelligent life is also possible in the case of Red Dwarfs, as the central star.

10.2 - The extended Seager-Equation

The Seager-Equation treats only a certain amount of Red Dwarf stars, i.e. stars of the spectral class **M**, namely only stars that can be detected by the JWST Telescope.
The Seager-Equation can also be extended to **all Red Dwarf** stars, in the galaxy.

N* = N$_{RD}$ = A·F$_{RD}$ is the number of M stars (Red Dwarfs) present in the galaxy with **F$_{RD}$** = 0.7 and **A** = 100 - 300 billion stars.

f_Q stands for the proportion of quiet M-stars, which account for about 80 %.

$f_{HZ} = F_{ph} = F_p \cdot F_h$ is the proportion of those M stars, which have a planet first and secondly it is in the habitable zone. F_p = 1:70 and F_h = 1:60 with F_{ph} = 1:4,200.

f_O quantifies the proportion of those planets that are visible at the star for the JWST. (0.01 x 0.1= 0.001)

$f_L = F_L$ is the fraction of animated planets, with F_L = 1:9.

f_S stands for an intelligence which leaves a measurable bio signature in the atmosphere, i.e. a technological civilization, with:
$f_S = F_i \cdot F_z$ = 1:14 · 1:7.943 = 1:111.203.

This results in:

10.2.1 Equation
$$N = A \cdot F_{RD} \cdot f_Q \cdot F_{ph} \cdot f_O \cdot F_L \cdot F_i \cdot F_z$$
$$N = A \cdot F_{RD} \cdot f_Q \cdot f_O \cdot F_{ph} \cdot F_{Liz}$$

Substituting the values into Equation 10.2.1 yields:

$N = (100 - 300) \cdot 10^9 \cdot 0.7 \cdot 0.8 \cdot 0.001 \cdot 1:4,200 \cdot 1:1,001$
N = 14 – 40   technological civilizations

According to Theorem 6.4.1, there are **10 – 290** technological civilizations, in Sun-like star systems, in our galaxy.
This results in a much **smaller** number of civilizations, with Red Dwarfs as central stars, than in Sun-like systems.

10.2.2 Equation $$N_S > N_{RD}$$

Comment:
A **bio signature** means certain signs in the atmosphere of an exoplanet, such as the existence of certain gases (such as CO_2 and CFCs) that indicate the presence of a technological civilization.

10.3 - The transformed Seager-Equation

1 to 7.

The Seager-Equation treats only Red Dwarfs, that is a certain set of stars, the so-called M-stars. One can extend the consideration here and also refer to other sets of stars, e.g. on the **G-stars** the Sun-like stars.
The Seager-Equation is then **compatible** with Equation System 6.3.3 and can be completely replaced by the relationships found in chapters

$N^* = N_s = A \cdot F_s$ according to Equation 1.4.1, for the number of G stars (Sun-like stars) present in the galaxy with F_s = 7:25 and A = 100 - 300 billion stars.

f_Q do not throw gamma rays into space, so all stars are observable, so $f_Q = 1$. The factor can therefore be dispensed with.

$f_{HZ} = F_{ph} = F_p \cdot F_h$ is the proportion of those G stars, which have a planet first, and secondly it is in a habitable zone. F_p = 201:14,000 and F_h = 10:603.

$f_O = F_k$ quantifies the proportion of those planets which visibly travel past the star for the Kepler telescope. According to chapter 1.2, the probability of such a transit is 0.47 % so F_k = 0.004,7.

$f_L = F_L$ is the fraction of planets with life, with F_L = 1:9.

f_S stands for an intelligence which leaves a measurable bio signature in the atmosphere, i.e. a technological civilization, with $f_S = F_i \cdot F_z$ = **1:14 · 1:7.943 = 1:111.203**.

The entire Seager-Equation can then be applied to the set of solar-like star systems in the galaxy observed with the Kepler telescope (or equivalent).
All the probability factors of the Seager-Equation are completely replaceable by the factors from the Equation System 6.3.3. The transformed Seager-Equation for G stars is then:

10.3.1 Equation $N = A \cdot F_s \cdot F_p \cdot F_h \cdot F_k \cdot F_L \cdot F_i \cdot F_z$

According to Definition 1.7.1 is: $F_{sph} = F_s \cdot F_p \cdot F_h = 1:15,000$

According to Definition 6.2.2 is: $F_{Liz} = F_L \cdot F_i \cdot F_z = 1:987$

Equation 10.3.1 can thus also be written as **transformed Seager-Equation**:

10.3.2 Equation $\boxed{N = A \cdot F_{sph} \cdot F_k \cdot F_{Liz}}$

Substituting all the values into Equation 10.3.2:

$N = (100 - 300) \cdot 10^9 \cdot 1:15,000 \cdot 0.004,7 \cdot 1:1,001$
N = 32 – 94   technological civilizations

Comparison Special Basic Model
Equivalent and thus comparable to the transformed Seager-Equation is Equation 6.3.3 from the Special Basic Model. According to Theorem 6.4.1 of the Special Basic Model, there are probably **10 - 290** technological civilizations, on „Earth 2" in solar-like systems, in our galaxy.
The Seager window is well located in the lower part of the Basic Model window.
The Drake-corrected Special Basic Model 9.8.2 delivers **22 - 199** „Earths 2" with technological civilizations. The Seager window is well located in the lower part of the Drake window.

10.3.3 Theorem   The Special Basic Model and the transformed Seager-Equation represent two mutually <u>equivalent</u> approaches.

In the Seager Approach, Earth-Similarity plays no role and only technological civilizations, on habitable planets in the galaxy, are asked.

This model can also be transferred to other star sets and observation devices. If you omit the factor F_z then you can apply Equation 10.3.2 to intelligent species. If you omit the factor F_i then Equation 10.3.2 can also be applied to **planets with life**.

11 – Equivalence of Considerations

11.1 - Equivalence

As the transformation shows overall, the Seager-Equation, after adaptation to Sun-like stars, can be regarded as equivalent to the Special Basic Model 6.3.3.
For this to be exactly the case, both equations must be equivalent to each other.

According to the Basic Model: $\quad N_{ze} = A \cdot F_{sph} \cdot F_{gae} \cdot F_{Liz}$

The Seager Approach: $\quad N = A \cdot F_{sph} \cdot F_{k} \cdot F_{Liz}$

It must then apply:

$$N_{ze} = N$$

$$A \cdot F_{sph} \cdot F_{gae} \cdot F_{Liz} = A \cdot F_{sph} \cdot F_{k} \cdot F_{Liz}$$

$$F_{gae} = F_{k}$$

If $F_{gae} = F_{g} \cdot F_{a} \cdot F_{e}$ is used according to Definition 3.1.3, then an assessment can be made for F_{e}. It turns out:

$$F_{gae} = F_{g} \cdot F_{a} \cdot F_{e} = F_{k}$$

Converting the equation to F_{e}:

11.1.1 Equation

$$\boxed{F_{e} = \frac{F_{k}}{F_{g} \cdot F_{a}}}$$

Substituting the values in Equation 11.1.1 yields:

F_{e} = 0.004,7 / (1,212:3,286 · 51:130)
F_{e} = 0.032,481 ≈ **1:31**

This is only three times greater than the previous minimum value, and can be used as a new upper limit. With the new borders, some theorems formulated so far can be differentiated somewhat.

11.2 - Corrected Values for the Earth

From the last chapter 11.1 there was a correction for the probability F_e of Earth-like, habitable planets:

11.2.1 Equation $\quad 1:100 \leq F_e \leq 1:31$

You can generate two values for F_{gae}:

F_{gae1} = 1,212:3,286 · 51:130 · 1:31 ≈ **1:214**
F_{gae2} = 1,212:3,286 · 51:130 · 1:100 ≈ **1:691**

A differentiated estimate for the probability factor F_{gae} can be given.

11.2.2 Equation $\quad 1:691 \leq F_{gae} \leq 1:214$

Equation 3.1.1 for habitable „Earth 2" in Sun-like systems is:

$$N_{he} = A \cdot F_{sph} \cdot F_{gae}$$

Insert the new values into Equation 3.1.1 yields:

N_{he} = (100 - 300)·10^9 · 1:15,000 · (1:214 − 1:691)

This results in the following correction for Theorem 3.1.3:

11.2.3 Theorem: **There are probably 9,648 to 93,458 „Earths 2", in solar-like star systems, in our galaxy.**

The Drake-corrected Special Basic Model 9.6.3 delivers **6,184 − 55,710** „Earths 2". The Seager window overlaps with the Drake window. Thus there is a good correspondence between the Drake-corrected Special Basic Model and the Seager-corrected Special Basic Model.

The probability of finding an „Earth 2" among habitable planets amounts to:

F_{gae2} = $F_g \cdot F_a \cdot F_{e2}$
F_{gae2} = 1,212:3,286 · 51:130 · 1:31
F_{gae2} = 0.004,667,67 ≈ **1:214**

This means that „Earth 2" can be found among **214 habitable planets**.
To find these habitable planets, you need to study 12,840 solar-like star systems that have planets. Thus 890,800 Sun-like star systems and a total of 3.21 million star systems would have to be observed and analysed. That is **21** times the amount of stars examined so far.

All cases summarized results in the following statement:

11.2.4 Theorem   **6.9-fold to 21-fold amount of stars that were examined with the Kepler telescope until 2013 are still needed to find a second Earth.**

Accordingly, it is expected that a second Earth will be found within the **next decades**.

11.3 - Corrections for Life and Intelligence

Equation 4.2.1 for habitable „Earths 2", with life in solar-like systems:

$$N_{Le} = A \cdot F_{sph} \cdot F_{gae} \cdot F_L$$

Insert the new values into Equation 4.2.1 yields:

$$N_{Le} = (100 - 300) \cdot 10^9 \cdot 1{:}15{,}000 \cdot (1{:}214 - 1{:}691) \cdot 1{:}9$$

This results in the following correction for Theorem 4.2.2:

11.3.1 Theorem: **There are probably 1,072 to 10,384 „Earths 2" with life in solar-like star systems in our galaxy.**

The Drake-corrected Special Basic Model 9.7.1 delivers **687 - 6,190** „Earths 2" with life. Thus the windows overlap and there is a good match between the Drake-corrected Special Basic Model and the Seager-corrected Special Basic Model.

Equation 5.3.1 for habitable „Earths 2" with intelligent species in Sun-like systems:

$$N_{ie} = A \cdot F_{sph} \cdot F_{gae} \cdot F_L \cdot F_i$$

Insert the new values into Equation 5.3.1 yields:

$N_{ie} = (100 - 300) \cdot 10^9 \cdot 1{:}15{,}000 \cdot (1{:}214 - 1{:}691) \cdot 1{:}9 \cdot 1{:}14$

This results in the following correction for Theorem 5.3.2:

11.3.2 Theorem: There are probably 77 to 742 „Earth 2" with intelligent species, in solar-like star systems, in our galaxy.

The Drake-corrected Special Basic Model 9.7.2 delivers **49 - 442** „Earths 2" with intelligent species. Thus the windows overlap and there is a good match between the Drake-corrected Special Basic Model and the Seager-corrected Special Basic Model.

11.4 - Corrections for the Basic Model

The **Special Basic Model** 6.3.3 for habitable „Earths 2", in solar-like systems, with a technological civilization is called:

$$N_{ze} = A \cdot F_{sph} \cdot F_{gae} \cdot F_{Liz}$$

Insert the new values into Equation 6.3.3:

$N_{ze} = (100 - 300) \cdot 10^9 \cdot 1{:}15{,}000 \cdot (1{:}214 - 1{:}691) \cdot 1{:}1{,}001$

This results in the following correction for Theorem 6.4.1:

11.4.1 Theorem There are probably 10 to 94 technological civilizations on an „Earth 2", in Sun-like star systems, in our galaxy.

Comparison Transformed Seager-Equation
Comparable here is the transformed Seager-Equation 10.3.2 with **32 - 94** technological civilizations. The two windows are approximately congruent and thus there is a good correlation of the Seager-Equation with the Seager-corrected Special Basic Model.

Comparison Drake-corrected Special Basic Model
The Drake-corrected Special Basic Model 9.8.1 delivers **6 - 56** „Earths 2" with technological civilizations. Thus the windows overlap and there is a good correlation between the Drake-corrected Special Basic Model and the Seager-corrected Special Basic Model.

The **General Basic Model** 8.4.2 for habitable „Earths 2", in the Galaxy, with a technological civilization is:

$$N_{zexGal} = \Sigma N_{zex} = A \cdot \Sigma(F_X \cdot F_{ph} \cdot F_{gae} \cdot F_{Liz})$$

$$N_{zexGal} \approx A \cdot F_{ph} \cdot F_{gae} \cdot F_{Liz}$$

Insert the new values into Equation 8.4.2:

N_{zexGal} = (100 - 300)·10^9 · 1:4,200 · (1:214 – 1:691) · 1:1,001

This results in the following correction for Theorem 8.4.5:

11.4.2 Theorem   **There are probably 10 to 94 technological civilizations on an „Earth 2", in solar-like star systems, in our galaxy.**

The corrected values inserted into the General Basic Model (as seen) is here referred as the **Seager-corrected General Basic Model**.

Comparison Drake-Equation
Comparable here is the Drake-Equation 9.1.2 with **9 - 109** extraterrestrial technological civilizations and thus lies well in the lower range of the Seager-corrected window.
The corrected Drake-Equation 9.5.3 provides **11 - 146** extraterrestrial technological civilizations and is thus well within the lower range of the Seager-corrected window.

Comparison Drake-corrected General Basic Model
The Drake-corrected General Basic Model 9.8.2 provides **22 - 199** „Earth 2" with technological civilizations. This overlaps the windows and there is a good match between the Drake-corrected General Basic Model and the Seager-corrected General Basic Model.
Thus, the **Drake-Equation** and the **Seager-corrected General Basic Model** show a good correlation.

11.5 - Other Civilizations

It is left to the reader to do the calculations for comparable, space travelling and old civilizations. See also the table on page 137, therefore only the results:

11.5.1 Theorem   The number of star systems in the galaxy, with Earth-like planets, in habitable zones that could support comparable civilizations, is at most likely between 26 and 252.

The General Basic Model 8.5.1 provides **26 - 781** comparable civilizations.
The Drake-corrected General Basic Model 9.9.1 provides **17 - 150** comparable civilizations.

11.5.2 Theorem   The set of stellar systems in the galaxy, with Earth-like planets, in habitable zones that could carry space-travelling civilizations, is most likely between 9 and 82.

The General Basic Model 8.5.2 provides **9 - 253** space travelling civilizations.
The Drake-corrected General Basic Model 9.9.2 provides **6 - 49** space travelling civilizations.

11.5.3 Theorem   The number of star systems in the galaxy, with Earth-like planets, in habitable zones that could support old civilizations, is at most likely between 2 and 14.

The General Basic Model 8.5.4 provides **2 - 43** old civilizations. The Drake-corrected General Basic Model 9.9.3 provides **1 - 8** old civilizations.

All Seager-corrected Basic Model windows are identical to the lower parts of the windows from the General Basic Model.
This ensures good agreement between the **General Basic Model** and the **Seager-corrected General Basic Model**.

Here too, the number of star systems from the General Basic Model is much larger than for the Drake-corrected data. It can be concluded that the upper bound F_e is set too high for the probability of a second Earth.

Furthermore, all Drake-corrected windows fit into the Seager-corrected windows. Thus, the **Drake-corrected Basic General Model** and the **Seager-corrected Basic General Model** are in good agreement.

11.6 - Result

There is a good agreement between the Seager-Equation and the Special Basic Model.
There is a good correspondence of the Seager-Equation with the Drake-corrected Special Basic Model.
This confirms that the **Special Basic Model** and the **transformed Seager-Equation** represent two **equivalent** approaches.

There is a good agreement between the Drake-Equation and the corrected Basic General Model.
This confirms that the **General Basic Model** and the **Drake-Equation** represent two **equivalent** approaches.

The compatibility of the **Drake-corrected Basic General Model** with the **Seager-corrected Basic General Model** is also evident.

11.6.1 Theorem The Special and General Basic Model, as well as the Drake-Equation and the Seager-Equation represent <u>equivalent</u> approaches.

12 – A General Approach

12.1 - Spectral Classes

In previous observations, star systems with a Sun-like central star were examined.
Assuming that other star systems, i.e. Non solar-like systems, also have distributions for technological civilizations, then the Seager-Equation 10.3.2 can be derived:

12.1.1 Equation $\quad N_X = A \cdot F_X \cdot F_{ph} \cdot F_k \cdot F_{Liz}$

Then applies to the star quantities formed by a solar type or **spectral class** [1] respectively.

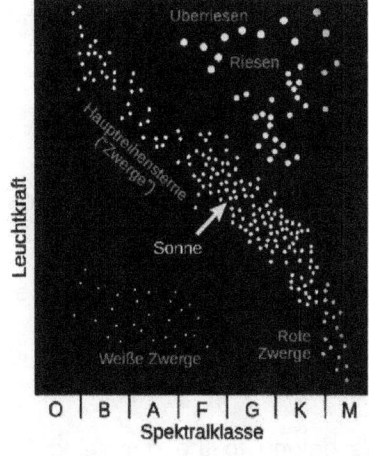

The suns of our galaxy are represented in the so-called **Hertzsprung-Russel-Diagram** [2] arranged according to colour and luminosity. A total of **13** spectral types exist.
In addition, there are the brown dwarfs and the red giants, i.e. the spectral classes **L, T, Y, R, N, S**, which also make up **1 %** of the total stars.
Two classes are known so far. This is the set of Sun-like **G** stars, with a yellow spectral colour and the probability $F_s = 0.28 = 7:25$.

As well as the number of Red Dwarfs, i.e. **M** stars with a red-orange spectral colour and a probability $F_{RD} = 0.7 = 7:10$. Thus, the two spectral classes make up **98 %** of the total stars in the galaxy.

12.2 - Civilizations in the Galaxy

N_X therefore represents the number of civilizations related to a particular type of sun. In order to determine the number of **all civilizations** in the galaxy, the sum of all partial results, i.e. across **all spectral classes**, must then be formed:

12.2.1 Equation $\quad N_{Civ} = \Sigma N_X = \Sigma (A \cdot F_X \cdot F_{ph} \cdot F_k \cdot F_{Liz})$

The number **A** of stars in the galaxy is the same for all star types and can therefore be drawn from the sum:

12.2.2 Equation $\quad \boxed{N_{CivGal} = \Sigma N_X = A \cdot \Sigma (F_X \cdot F_{ph} \cdot F_k \cdot F_{Liz})}$

The probabilities F_X, F_{ph}, F_{Liz} depend on the spectral class of the respective star set and F_k on the observation instrument used. All factors can be empirically determined in the long run, according to Theorem 6.1.2 within 2 centuries.

Based on the Seager Approach, Equation 12.2.2 represents the most general form in which the subject of intelligent life or technological civilizations, on a habitable planet, in the galaxy, can be mathematically represented.

Equation 12.2.2 is therefore referred as the „General Approach". Equation 12.2.2 can be used to estimate the number of civilizations in the galaxy.

According to previous observations, the distribution for **G** stars, i.e. Sun-like stars, is about 28 % of the total stars.
The Red Dwarfs, i.e. the **M** stars, represent the major part, with about 70 % of the total stars, of which only 80 % are observable.
Therefore: $N^* = A \cdot F_{RD} \cdot f_Q$.
The remaining 2 % of the total stars belong to the remaining **11** spectral classes and are not taken into account in this estimation without impairing the observation.

The probabilities for Sun-like stars are known according to chapters 1 to 7. The data for the Red Dwarfs are only partly taken from the data of Sara Seager, because some of her assumptions (life and biosphere) are too optimistic. Then the following values are used for Equation 12.2.2:

$$N = A \cdot F_{sph} \cdot F_k \cdot F_{Liz}$$
$$+ A \cdot F_{RD} \cdot f_Q \cdot F_{ph} \cdot f_O \cdot F_{Liz}$$

$$N = (100 - 300) \cdot 10^9 \cdot 1:15,000 \cdot 0.004,7 \cdot 1:1,001$$
$$+ (100 - 300) \cdot 10^9 \cdot 0.7 \cdot 0.8 \cdot 1:4,200 \cdot 0.001 \cdot 1:1,001$$

$$N = 32 - 94$$
$$+ 14 - 40$$

$$N = 46 - 134 \quad \text{technological civilizations}$$

Comparison General Basic Model
According to Theorem 8.4.5 of the General Basic Model, the number of star systems, with Earth-like planets, in habitable zones that could support civilizations is probably between **35 - maximum 1,034**.
The Drake-corrected General Basic Model 9.8.2 provides **22 - 199** „Earths 2" with technological civilizations.
The Seager-corrected General Basic Model 11.4.2 provides **35 - 334** „Earths 2" with technological civilizations.
The calculated window of the General Approach corresponds well with the window for the General Basic Model and thus results in a good match between the two approaches.
This results in a good agreement between the **General Approach** and the previous probability considerations from the **corrected General Basic Model**.
This confirms that the **General Approach** and the **General Basic Model** are two **equivalent** approaches.

Comparison Drake-Equation
The Drake-Equation 9.1.2 provides **9 - 109** extraterrestrial technological civilizations and is thus well within the range of the calculated window of the General Approach.
The corrected Drake-Equation 9.5.3 lets expect **11 - 146** extraterrestrial technological civilizations and is thus well within the range of the calculated window of the General Approach.
This results in a good agreement of the **General Approach** with the **Drake-Equation**.
This confirms that the **General Approach** and the **Drake-Equation** represent two equivalent approaches.

12.3 - Technological Civilizations

For an approximate maximum calculation of the number of technological civilizations in the galaxy, it can be assumed that the probability for a habitable planet F_{ph} and also the probability for civilizations F_{Liz} are the same for all star types.

12.3.1 Approach

In star systems that are Non Sun-like, there are probably the same or similar distributions, for habitable planets, with civilizations, as in Sun-like systems.

The factors F_{ph} and F_{Liz} can then be drawn from the sum of Equation 12.2.2.
As long as the observation is done with a single instrument, F_k does not change and can therefore also be drawn from the sum formation. It then applies:

$$N_{Civ} = \Sigma N_x \approx A \cdot F_{ph} \cdot F_k \cdot F_{Liz} \cdot \Sigma(F_x)$$

The F_x represent the distribution of the individual spectral classes. The sum, of all single distributions of the spectral classes, results again in the totality of the stars, in the galaxy. Therefore:

$$\Sigma(F_x) = 1$$

It results approximately in total:

12.3.2 Equation

$$N_{Civ} = \Sigma N_x \approx A \cdot F_{ph} \cdot F_k \cdot F_{Liz}$$

For F_k you can use two values, the G-star and the M-star.

Inserting the values ($F_k = 0.004,7$) into Equation 12.3.2:

$N_{Civ1} \approx (100 - 300) \cdot 10^9 \cdot 1{:}4{,}200 \cdot 0.004{,}7 \cdot 1{:}1{,}001$
$N_{Civ1} \approx \mathbf{112 - 336}$ **technological civilizations**

Inserting the values ($F_k = 0.001$) into Equation 12.3.2:

$N_{Civ2} \approx (100 - 300) \cdot 10^9 \cdot 1:4{,}200 \cdot 0.001 \cdot 1:1{,}001$
$N_{Civ2} \approx 24 - 72$ technological civilizations

12.3.3 Theorem   There are probably a maximum of 24 to 336 technological civilizations in our galaxy.

The same considerations apply to the General Basic Model and the Drake-Equation as in the previous chapter 12.2.

12.3.4 Theorem   The General Approach and the General Basic Model, as well as the Drake-Equation represent <u>equivalent</u> approaches to each other.

If the factor F_z is omitted, Equation 12.2.2 can be applied to **intelligent species**. If you also omit the factor F_i then Equation 12.2.2 can also be applied to **animated planets**.

12.4 - Other Civilizations

It is left to the reader to make the calculations for comparable, space travelling and old civilizations himself. See also the table on page 137, so here are the results:

12.4.1 Theorem   There are probably a maximum of 18 to 254 comparable civilizations in our galaxy.

The General Basic Model 8.5.1 provides **26 - 781** comparable civilizations.
The Drake-corrected General Basic Model 9.9.1 provides **17 - 150** comparable civilizations.
The Seager-corrected General Basic Model 11.5.1 provides **26 - 252** comparable civilizations.

12.4.2 Theorem   There is probably a maximum of 6 to 82 space travelling civilizations in our galaxy.

The General Basic Model 8.5.2 provides **9 - 253** space travelling civilizations.
The Drake-corrected General Basic Model 9.9.2 provides **6 - 49** space travelling civilizations.
The Seager-corrected General Basic Model 11.5.2 provides **9 - 82**

space travelling civilizations.

12.4.3 Theorem   There are probably no more than 1 to 14 old civilizations in our galaxy.

The General Basic Model 8.5.4 provides **2 - 43** old civilizations.
The Drake-corrected General Basic Model 9.9.3 provides **1 - 8** old civilizations.
The Seager-corrected General Basic Model 11.5.3 provides **2 - 14** old civilizations.
The General Approach always matches the values of the General Basic Model.
The General Approach also matches the values of the Drake- and the Seager-corrected General Basic Model in all cases.
This confirms once again that the **General Approach** and the **General Basic Model** represent two **equivalent** approaches.

12.5 – Basic Model and General Approach

As chapter 12 shows, the General Basic Model can be regarded as equivalent to the General Approach 12.2.2.
For this to be exactly the case, both equations must be equivalent to each other.

General Approach: $N_{Civ} = A \cdot F_{ph} \cdot F_k \cdot F_{Liz}$

General Basic Model: $N_{zexGal} = A \cdot F_{ph} \cdot F_{gae} \cdot F_{Liz}$

It must then apply:

$$N_{Civ} = N_{zexGal}$$

$$A \cdot F_{ph} \cdot F_k \cdot F_{Liz} = A \cdot F_{ph} \cdot F_{gae} \cdot F_{Liz}$$

12.5.1 Equation $F_k = F_{gae}$

The probability F_{gae} for the Earth Similarity must be of the **same order of magnitude** as the observation probabilities F_k, so that the models are equivalent.

12.6 - Drake-Equation and General Approach

As the transformation shows, after adaptation to the Basic Model, the Drake-Equation can be regarded as equivalent to the General Approach 12.2.2.
For this to be the case, both equations must be equivalent to each other.

General Approach: $\quad N_{Civ} = A \cdot F_{ph} \cdot F_k \cdot F_{Liz}$

The transformed Drake-Equation: $\quad N = R \cdot F_p \cdot F_{Liz} \cdot L$

It must then apply:
$$N_{Civ} = N$$

$$A \cdot F_{ph} \cdot F_k \cdot F_{Liz} = R \cdot F_p \cdot F_{Liz} \cdot L$$

$$A \cdot F_h \cdot F_k = R \cdot L$$

From the last chapter 12.4 applies: $F_k = F_{gae}$ Insertion into the equation delivers:
$$A \cdot F_h \cdot F_{gae} = R \cdot L$$

Resolving the equation after F_{gae} yields:

$$F_{gae} = R \cdot L / (A \cdot F_h)$$

Further dissolution into F_e yields:
$$F_e = R \cdot L / (A \cdot F_h \cdot F_g \cdot F_a)$$

For the probability of an Earth F_e results with:

$F_e = 535{,}461 \cdot 11.5 / (100 - 300) \cdot 10^9 \cdot 10{:}603 \cdot 1{,}212{:}3{,}286 \cdot 51{:}130$
$F_e = 0.008{,}553 - 0.025{,}661$
$F_e = 1{:}117 - 1{:}39$

For the similarity of the Earth F_{gae} this results in:

$F_{gae} = F_g \cdot F_a \cdot F_e$
$F_{gae} = 1{,}212{:}3{,}286 \cdot 51{:}130 \cdot (1{:}117\text{-}1{:}39)$
$F_{gae} = 1{:}808 - 1{:}269$

12.7 - Corrections for the Earth

The General Approach thus provides a total:

$$1:117 \leq F_e \leq 1:39$$
$$1:808 \leq F_{gae} \leq 1:269$$

The Drake-corrected Special Basic Model 9.6.1 provides:

$$1:156 \leq F_e \leq 1:52$$
$$1:1,078 \leq F_{gae} \leq 1:359$$

The Seager-corrected Special Basic Model 11.2.1 provides:

$$1:100 \leq F_e \leq 1:31$$
$$1:691 \leq F_{gae} \leq 1:214$$

This makes it possible to determine the maximum limits for the probability of a second Earth:

12.7.1 Equation $\boxed{1:152 \leq F_e \leq 1:31}$

This results in the Earth's Similarity:

12.7.2 Equation $\boxed{1:1,078 \leq F_{gae} \leq 1:214}$

The 21 to 108 times the number of stars that have been studied with the Kepler telescope by 2013 is still needed to find an „Earth 2".
After the result, it will be **another decade or two** before a second Earth is found. If you take the current speed of discovery as a basis. It would therefore be a pure stroke of luck if a second Earth were to be found in the next few years.
A faster discovery of a second Earth is only possible if, in a shorter period of time, even more stars were examined than have been observed to date.

The Basic Model corrected by Equations 12.7.1 and 12.7.2 is referred as the **corrected Special** or **General Basic Model**.

13 – Lines of Evolution

13.1 - Lines of Development on the Earth

What would have happened if 65 million years ago no asteroid had fallen on the Earth and the dinosaurs would have had time to develop? Some scientists are of the opinion that the dinosaurs were so successful in evolutionary biology that they had occupied all niches in the ecosystem and that there was no room for further evolution.
However, recent findings indicate that under the Theropods, [1] especially the Troodon [2] a development of the social intelligence took place. In the case of undisturbed development, a genuine intelligence with self-consciousness could have developed there.
Dale Alan Russell, a Canadian palaeontologist dealing with dinosaurs, speculated about a hypothetical intelligent end product of the dinosaur evolution. [3]
Man has developed in 65 million years from a small mouse-sized mammal to Homo sapiens. [4] This suggests that there was also a chance for the dinosaurs to develop intelligence. And what would have become of the evolutionary lines that have been eliminated by other catastrophes? Let's look at the interrupted development possibilities:

1) About 80 % of all species of animals and plants, including the trilobites (trilobites), but also conodonts or brachiopods (bony feet), died about 485 million years ago, at the end of the **Cambrian Period** [5]. The insects spread. [6]

2) About 360 million years ago, in the **upper Devon** (Kellwasser Event), [7] about 50 % of all species died, including some fish, corals and trilobites. Then came the age of amphibians. [6]

3) About 252 million years ago, within a period of 200,000 years at the **Perm-Trias Border**, [8] 95 % of all sea-dwelling species, as well as about 66 % of all land-living species (reptile and amphibian species) died. The age of the Therapsids began. These are mammalian-like reptiles. [6]

4) About 200 million years ago, at the end of the **Triassic Period** [9], 50 to 80 % of all species, including almost all farm animal animals, died. It was followed by the age of dinosaurs. [6]

5) Approximately 66 million years ago, on the **Cretaceous Tertiary Border**, [10] around 50 % of all animal species died, including the dinosaurs. It began the age of the mammals, from which we developed. [6]

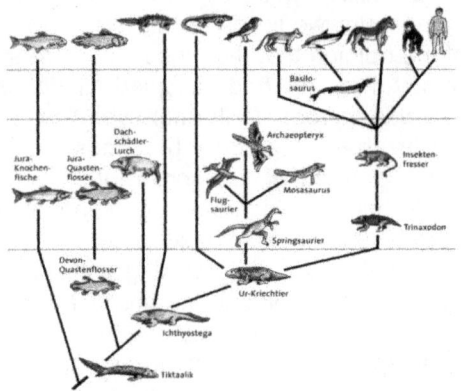

It is conceivable that all four interrupted evolutionary strands could have developed further in such a way that intelligent species would have emerged from them.

Therefore the following axiom is set up:

13.1.1 Axiom   Every evolutionary line, with undisturbed development, can produce life forms that are conscious and intelligent.

As the development of mammals was the most recent development on Earth, the number **N** of lizard-based civilizations is greater than mammalian. And that e.g. more reptiloid species than a sauroid species. The following evolutionary strings are conceivable:

1) Insects
2) Amphibious
3) Reptiloids
4) Sauroids
5) Humanoids

13.1.2 Assumption

$$N_{insects} \geq N_{amphibious} \geq N_{reptiloid} \geq N_{sauroid} \geq N_{humanoid}$$

13.2 - Convergent Development

The biological convergence theory [11] is based on the assumption that many similar functionalities have arisen in evolution, independently of one another, through the functional constraints prevailing on the Earth. Examples are the wings or the eye, both of which have been independently developed by several species.
Simon Conway Morris [12] is a British palaeontologist. He is the principal representative of the theory of convergence in evolution and the opinion that life is stable because nature has provided the framework for it, and life inevitably follows the selective-adaptive rules. Therefore evolution must necessarily arrive at an intelligent species. The development to complexity and intelligence is almost **a program part within evolution.**

The premise of biochemistry and biophysics, based on the periodic system of the elements and also the "codon triplet guide" for building functional units on protein structures, can be formulated as follows:

13.2.1 Axiom   DNA is a universal blueprint for the development of a species.

The development to complexity and intelligence would have to take place on every planet on which this is possible.

Furthermore, it may be assumed that there is a certain similarity even in the case of different developmental strands. All species will thus have the following body parts:

1) left-right symmetry
2) Fuselage for respiratory and digestive organs
3) upper extremities for handling objects
4) lower extremities for movement
5) Head with sensor organs

So is to be assumed that **most** intelligent species have developed a **humanoid-like** or **humanoid-analog** shape.

13.3 - Humanoids in Sun-like Systems

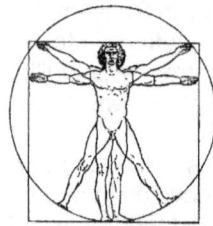

Besides the 5 evolutionary strands mentioned above, the sixth possibility would be a completely different structured development, e.g. an evolutionary strand that has formed an exoskeleton like insects. The chance of encountering a humanoid species is therefore 1 to 7, so the probability factor is $F_m = 0.142,8 = 1:7$.

Derived from Equation System 6.3.3, the number of humanoid species in the galaxy is:

13.3.1 Equation $N_{me} = A \cdot F_{sph} \cdot F_{gae} \cdot F_{Liz} \cdot F_m$

Inserting all values (F_{gae} = 1:214) into Equation 13.3.1:

$N_{me1} = (100 - 300) \cdot 10^9 \cdot 1:15,000 \cdot 1:214 \cdot 1:1,001 \cdot 1:7$
$N_{me1} = 5 - 14$ **human civilizations**

Inserting all values (F_{gae} = 1:1,078) into Equation 13.3.1:

$N_{me2} = (100 - 300) \cdot 10^9 \cdot 1:15,000 \cdot 1:1,078 \cdot 1:1,001 \cdot 1:7$
$N_{me2} = 1 - 3$ **human civilizations**

The two results can then be summarized to the following statement:

13.3.2 Theorem   **There could be between 1 to 14 human-like species, in solar-like star systems, in the galaxy.**

The other species would be different in form, in biology, in biochemistry and genetics, and also different in civilisation and social understanding.
At worst case, we could be the only human species in a Sun-like system.

The probability of a really humanoid species is:

13.3.3 Definition $F_{me} = F_{sph} \cdot F_{gae} \cdot F_{Liz} \cdot F_m$

F_{me} = 1:15,000 · (1:1,078-1:214) · 1:1,001 · 1:7
F_{me} = 1:88,258,768,350 – 1:444,593,235,000

Only every **88.258 - 444.593 billionth** Sun-like system could then accommodate a truly human-like species.

13.4 - Corrected General Basic Model

If one assumes that stars that are Non Sun-like also have an equal or similar distribution for humanoid species, then the General Basic Model can be applied.

13.4.1 Approach   In star systems that are Non Sun-like, there are probably the same or similar distributions for humanoid species as in Sun-like systems.

Equation 8.4.1 can then be modified:

13.4.2 Equation $N_{mex} = A \cdot F_X \cdot F_{ph} \cdot F_{gae} \cdot F_{Liz} \cdot F_m$

The number of N_{mexGal} of all star systems, with habitable planets similar to Earth and with human-like species in the galaxy, is then approximated ($\Sigma F_x = 1$):

13.4.3 Equation $N_{mexGal} = \Sigma N_{mex} \approx A \cdot F_{ph} \cdot F_{gae} \cdot F_{Liz} \cdot F_m$

Inserting all values ($F_{gae} = 1:214$) into Equation 13.4.3:

$N_{mexGal1} \approx (100 - 300) \cdot 10^9 \cdot 1:4,200 \cdot 1:214 \cdot 1.1,001 \cdot 1:7$
$N_{mexGal1} \approx$ **16 – 48**   humanoid species in the galaxy

Inserting all values ($F_{gae} = 1:1,078$) into Equation 13.4.3:

$N_{mexGal2} \approx (100 - 300) \cdot 10^9 \cdot 1:4,200 \cdot 1:1,078 \cdot 1:1,001 \cdot 1:7$
$N_{mexGal2} \approx$ **3 – 10**   humanoid species in the galaxy

The two results can then be summarized to the following statement:

13.4.4 Theorem   There could be between **3 to 48 humanoid species in the galaxy.**

According to Theorem 8.4.5, the number of star systems, with Earth-like planets, in habitable zones that civilizations can support is probably between **35 - 1,034** at most, and the number of humanoid civilizations is only a small part of the existing civilizations in the galaxy.

13.4.5 Equation $\quad\boxed{N_{zeGal} > N_{mzGal}}$

The probability of finding an Earth-like planet with a humanoid species is then:

13.4.6 Definition $\quad F_{mex} = F_{ph} \cdot F_{gae} \cdot F_{Liz} \cdot F_m$

F_{mex} = 1:4,200 · (1:1,078-1:214) · 1:1,001 · 1:7
F_{mex} = 1:315,209,887,000 – 1:15,878,329,820

Only every **15.878 - 315.209 billion** star system has a planet, with a humanoid civilization.

13.5 - General Approach

The General Approach can also be used here. Equation 12.2.2 can then be modified:

13.5.1 Equation $\quad N_{mex} = A \cdot F_X \cdot F_{ph} \cdot F_k \cdot F_{Liz} \cdot F_m$

The number of N_{mexGal} of all star systems, with habitable planets similar to Earth and with human-like species in the galaxy, is then approximated ($\Sigma F_x = 1$):

13.5.2 Equation $\quad N_{mexGal} = \Sigma N_{mex} \approx A \cdot F_{ph} \cdot F_k \cdot F_{Liz} \cdot F_m$

For F_k you can use two values, the G-stars and the M-stars.

Inserting all values (F_k = 0.004,7) into Equation 13.5.2:

$N_{mexGal1} \approx (100 - 300) \cdot 10^9 \cdot 1:4,200 \cdot 0.004,7 \cdot 1:1,001 \cdot 1:7$
$N_{mexGal1} \approx 16 – 48 \quad$ **humanoid species in the galaxy**

Inserting all values (F_k = 0.001) into Equation 13.5.2:

$N_{mexGal2} \approx (100 - 300) \cdot 10^9 \cdot 1:4,200 \cdot 0.001 \cdot 1:1,001 \cdot 1:7$
$N_{mexGal2} \approx 4 – 10 \quad$ **humanoid species in the galaxy**

The two results can then be summarized to the following statement:

13.5.3 Theorem   **There could be between 4 to 48 humanoid or humanoid analog species in the galaxy.**

According to the General Basic Model 13.4.4, **3 - 48** humanoid species exist in the galaxy. Thus, there is agreement here in the General Basic Model and in the General Approach.

13.5.4 Theorem   **There is a good chance that we are not the only humanoids in the galaxy.**

13.6 - Working Hypothesis

All the assumptions and axioms made so far can be considered as **first assumptions** or **first approaches**, such as statements about life or evolution, or the same probabilities in Non Sun-like star systems, etc. For more precise future data, these assumptions can be further differentiated, depending on requirements and possibilities.

As with any falsifiable model, newer data can be used to modify or even revise individual statements, although the overall structure and also the content of the present model remain valid.
With all equations, definitions, probability factors, axioms and propositions available to date, an approach is now available which is based on experimental and empirical data, allows differentiations at any time through future data and can therefore serve as a **working hypothesis** for probability considerations of extraterrestrial civilizations.

If, through future investigations, an ever better significance of the previous probability values is achieved (according to Theorem 6.1.2 within the next two centuries), the probabilities of probability factors change into simple distribution or frequency values.
This also transforms the probability model into a simple, empirical distribution model of planets, life, intelligence and civilizations in our galaxy.

13.7 - Probabilities

Here again the values for all found probabilities factors:

Symbol	Rate	Factor	Designation
F_s	7:25	0.28	G-stars
F_{RD}	7:10	0.7	M-stars
F_p	201:14,000	0.014,357	G-stars with planets
F_h	10:603	0.016,583	planets in habitable zones
F_g	1,212:3,286	0.368,837	approximately Earth-great planets
F_a	51:130	0.392,307	approximately Earth-like planets
F_e	1:152 – 1:31	0.006 – 0.032	Earth-like planets
F_{hsub}	1:53	0.018,867	Subearth
F_{hsup}	2,043:3,286	0.621,728	Superearth
F_{sph}	1:15,000	0.000,066	habitable planets, G-star
F_{ph}	1:4,200	0.000,238	habitable planets
F_{gae}	1:1,078–1:214	0.0009-0.004	Earth Similarity
F_L	1:9	0.111...	planets with life
F_i	1:14	0.071,428	intelligent species
F_z	1:7.943	0.125,895	technological civilizations
F_z	1:10.522	0.095,038	comparable civilizations
F_z	1:32.448	0.030,817	space travelling civilizations
F_u	1:6	0.166...	old civilizations
F_m	1:7	0.142,857	humanoid civilizations
F_{Liz}	1:1,001	0.000,999	technological civilizations
F_{Liz}	1:1,326	0.000,754	comparable civilizations
F_{Liz}	1:4,089	0.000,244	space travelling civilizations
F_k	1:213	0.004,7	observability G-Stars
F_k	1:1000	0.001	observability M-Stars

14 – Approximately Earth-great planets

14.1 - Influence of Gravity

Subearths and superearths may produce lower life. The probability for intelligent species or even civilizations, however, is close to zero.
Apart from the Earth-like planets, only **approximately Earth-great planets** can be considered.
It should also be pointed out that moons, with the corresponding conditions, life, intelligence and civilization can also arise, e.g. as the moon of a giant planet in a habitable zone. This topic is not dealt with here, as too little data is still available.
In principle, **life, intelligence** and **civilization** can develop on planets about the size of the Earth, in solar-like star systems, as well as on planets similar to Earth. The Basic Model can also be used here.
According to the „*Habitable Exoplanets Catalog*" of December 2017, 21 of 53 habitable planets are about the size of the Earth. 17 of the found planets, about the size of the Earth, lie in a range of 1.3 to 4.8 Earth masses. 4 planets are outside this range.
Then there is a probability of F_{gg} = **17:21** for planets that could possibly carry life. Then the probability of an approximately Earth-great planet applies:

14.1.1 Equation

$F_{hgg} = F_g \cdot F_{gg}$
$F_{hgg} = 1,212:3,286 \cdot 17:21$
$F_{hgg} = 20,604:69,006$

It can be assumed that the probability of living with gravity decreases. An approach can be achieved by using the square of the Earth's masses to calculate the probability F_L. For all planets about the size of the Earth then applies:

$F_{Lg} = 1:m_E^2$ and m_E = number of Earth masses

In the following illustration on the next page, the connection is shown again graphically.

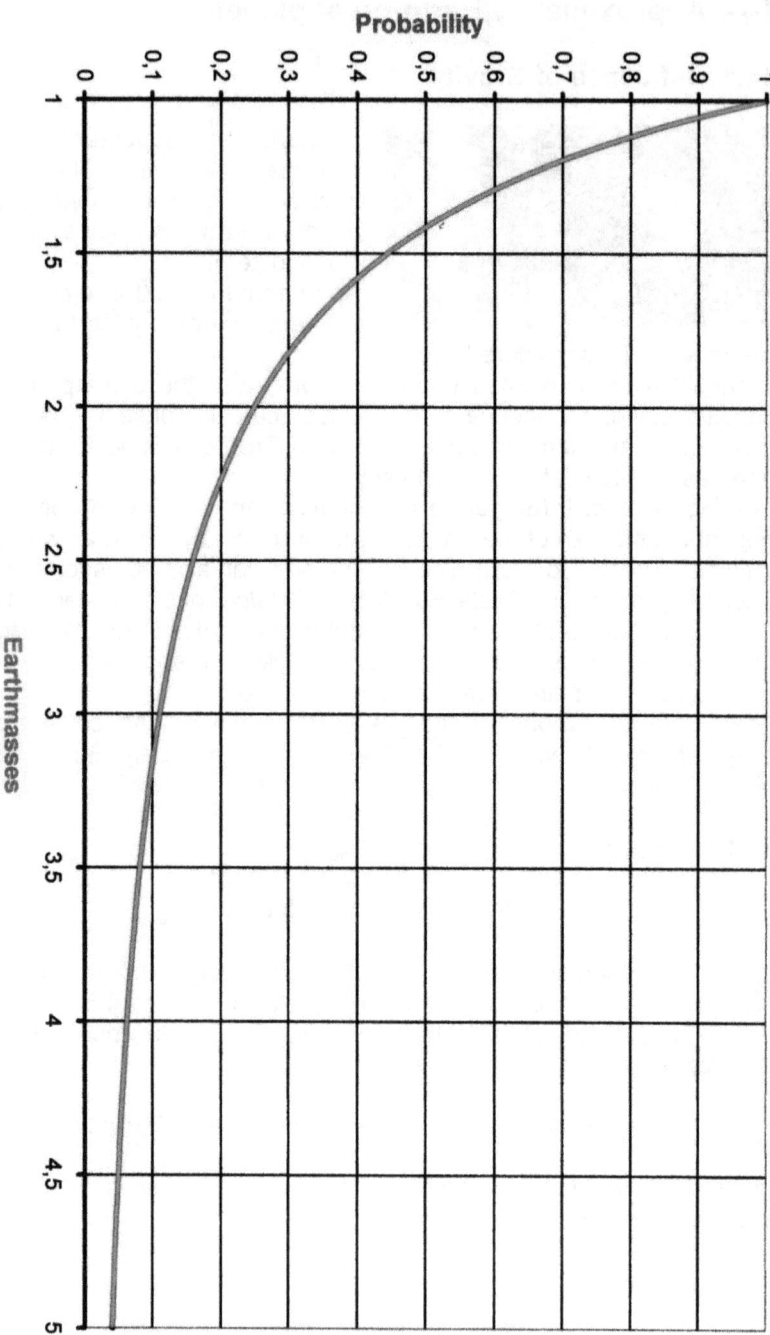

The evaluation of the planetary data is shown in the following table:

Number n	m_E Earth masses	Mean Value m_E	F_{Lg}
5	1.3 – 1.8	1.44	0.482,253
8	2.0 – 2.8	2.6	0.147,928
2	3.3 – 3.8	3.65	0.075,060
2	4.3 – 4.8	4.55	0.048,303

The probability for a group of approximately Earth-great planets then arises:

14.1.2 Equation $\quad F_{Lgm} = F_n \cdot F_{Lg} \quad$ and $\quad F_n = n:17$

For the probability, for all planets about the size of the Earth, the mean value results with:

14.1.3 Equation $\quad F_{LG} = (\Sigma F_{Lgm})/17$

$$\begin{aligned}
F_{LG} &= (5:17 \cdot 0.482,253 \\
&+ 8:17 \cdot 0.147,928 \\
&+ 2:17 \cdot 0.075,060 \\
&+ 2:17 \cdot 0.048,303)/17 \\
&= 0.225,965 /17 \\
F_{LG} &\approx 1:75
\end{aligned}$$

14.2 - Life and Civilization

According to chapter 4, the probability for an animate planet $F_L = 1:9$. and the total probability for animate planets about the size of the Earth:

14.2.1 Equation
$$\begin{aligned}
F_{Lgg} &= F_L \cdot F_{LG} \\
F_{Lgg} &= 1:9 \cdot 1:75 \\
F_{Lgg} &= 1:675
\end{aligned}$$

According to chapter 5, the probability for a living planet with an intelligent species is $F_i = 1:14$.
There is no reason why the same probability applies to approximately Earth-great planets.

As in chapter 6.2, the number of civilizations is expected to decrease across the development levels, using a sharpened approach.

14.2.2 Approach The probability F_z for a civilization is inversely proportional to the square of the development level.

$$F_z = 1:m^2 \quad \text{and} \quad m = \text{development level}$$

In the following graphic, on the next page, this is illustrated once again.

Only three of the eight levels of development represent higher technological civilizations, namely levels 6, 7 and 8, which corresponds to a probability of:

F_{zg} = 1:36 + 1:49 + 1:64
F_{zg} = 1,801:28,224
F_{zg} = 0.063,81 ≈ **1:16**

Overall, for the number of technological civilizations on approximately Earth-great planets:

14.2.3 Equation $N_{zg} = A \cdot F_{sph} \cdot F_{hgg} \cdot F_{Lgg} \cdot F_i \cdot F_{zg}$

Inserting all determined values into Equation 14.2.3 yields:

$N_{zg} = (100 - 300) \cdot 10^9 \cdot 1{:}15{,}000 \cdot 20{,}604{:}69{,}006 \cdot 1{:}675 \cdot 1{:}14 \cdot 1{:}16$

N_{zg} = 13 – 40 technological civilizations

Despite small probabilities, there are still a small number of possible technological civilizations.

Equation 14.2.3 is a **modification** of Equation 6.3.3 for Earth-like planets in Sun-like star systems. Here shown modified for habitable, approximately Earth-great planets, in solar-like star systems.

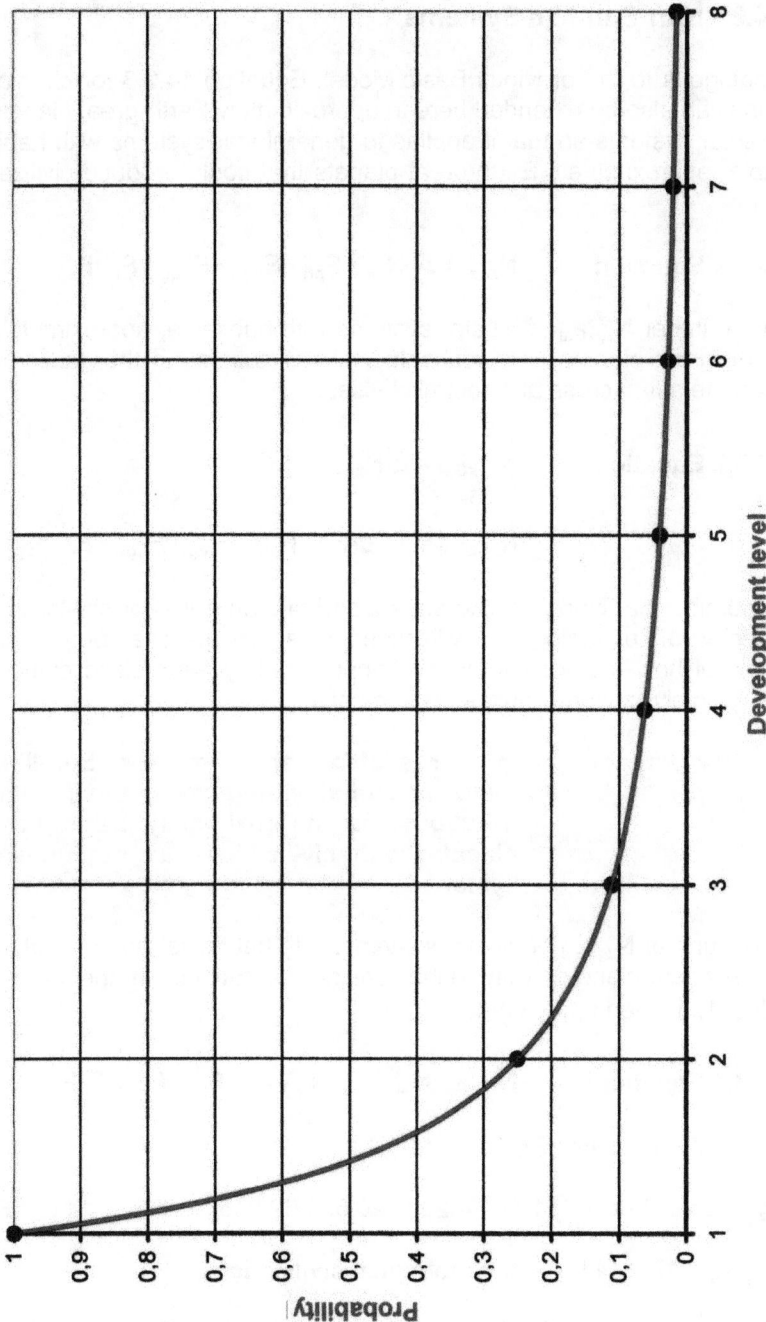

14.3 - Non Sun-like Systems

Analogous to the previous Basic Model, Equation 14.2.3 for civilizations can also be extended here to approximately Earth-great planets in solar systems so that it applies to general star systems with habitable, approximately Earth-great planets that could produce civilizations.

14.3.1 Equation $N_{zgx} = A \cdot F_x \cdot F_{ph} \cdot F_{hgg} \cdot F_{Lgg} \cdot F_i \cdot F_{zg}$

The number N_{zgxGal} of all star systems with habitable, approximately Earth-great planets, with technological civilizations, in the galaxy is then the sum across all spectral classes:

14.3.2 Equation $N_{zgxGal} = \Sigma N_{zgx}$

$$N_{zgxGal} = A \cdot \Sigma(F_x \cdot F_{ph} \cdot F_{hgg} \cdot F_{Lgg} \cdot F_i \cdot F_{zg})$$

Assuming that Non Sun-like stars also have an equal or similar frequency for technological civilizations, one can make a rough estimate of how many systems with approximately earth-sized planets carrying civilizations could exist at most.

14.3.3 Approach In star systems that are Non Sun-like, there are probably the same or similar distributions, for approximately Earth-great planets, with civilizations as in Sun-like systems.

The number N_{zgxGal} of all star systems, with habitable, approximately earth-sized planets, with technological civilizations, in the galaxy ($\Sigma F_x = 1$), is then approximated:

14.3.4 Equation $N_{zgxGal} = \Sigma N_{zex} \approx A \cdot F_{ph} \cdot F_{hgg} \cdot F_{Lgg} \cdot F_i \cdot F_{zg}$

Insert all values into Equation 14.3.4:

$N_{zgxGal} = (100 - 300) \cdot 10^9 \cdot 1{:}4{,}200 \cdot 20{,}604{:}69{,}006 \cdot 1{:}675 \cdot 1{:}14 \cdot 1{:}16$

$N_{zgxGal} = 47 - 141$ **technological civilizations**

These are more civilizations as in systems with Red Dwarfs, but less than civilizations on „Earth 2" in Sun-like systems.

Equation 14.3.4 is a modification of Equation 8.4.2 for civilizations on habitable planets in any star system in the galaxy. Shown here modified for habitable approximately Earth-great planets.

14.4 - Result

Overall, from the considerations in this chapter, there is a good probability that life, intelligence and civilization can also exist on **approximately Earth-great planets**.
Due to gravity, however, life there may take on different forms than we are accustomed to from the Earth.

With the approximately Earth-great planets from this chapter and the reflections on the Red Dwarfs from chapter 10, there is a good probability that life, intelligence and civilization could also arise on other planets or star systems.

Almost all possibilities for life, intelligence and civilization are covered by the approximately Earth-great planets around G-stars and M-stars.
Only two planetary systems and moons (orbiting giant planets in a habitable zone) remain that could produce life or civilization. However, no information is available here yet.

15 – Distributions

The values presented below are a summary of the calculations shown so far.

15.1 - Distribution of Raw Materials

Drake-corrected Special Basic Model
The distribution of raw materials on an „Earth 2" in the galaxy can be determined with the corrected values of the Earth.
According to Theorem 9.6.3, **6,184 - 55,710** „Earths 2" exist in our galaxy.
According to Theorem 9.7.1, the number of solar-like star systems, in the galaxy, with an „Earth 2" in habitable zone carrying life, is **678 - 6,190** systems.
If 55,710 „Earths 2" exist and only 6,190 of them carry life, then **49,520** inanimate, Earth-like planets remain, from which inorganic raw materials would have to be obtained.
According to Theorem 9.7.2, the number of solar-like star systems, in the galaxy, with an „Earth 2" in habitable zone, home to **49 - 442** intelligent species.
If 6,190 „Earths 2" exist with life and only 442 of them carry intelligence, then there are still **5,748** animate planets similar to Earth from which organic raw materials could be obtained.

Seager-corrected Special Basic Model
The distribution of resources on a „Earth 2" in the galaxy can be determined with the corrected values of the Earth.
By Theorem 11.2.3 there are **9,648 - 93,458** „Earths 2" in our galaxy.
By Theorem 11.3.1, the number of Sun-like stellar systems in the galaxy, with a „Earth 2" in a habitable zone that sustains life, is **1,072 - 10,384** systems.
If 93,458 „Earths 2" exist and only 10,384 of them live, then there are still **83,074** inanimate Earth-like planets from which to extract inorganic resources.
By Theorem 11.3.2, the number of solar-like star systems in the galaxy, with a „Earth 2" in a habitable zone that is home to intelligent species, is **77 - 742** systems.
If 10,384 „Earths 2" exist with life and only 742 of them carry intelligence, then there are still **9,642** live, Earth-like planets, from which organic resources could be fetched.

Therefore, the overall result of the corrected Basic Model is:

15.1.1 Theorem There could be between 49,520 to 83,074 animated Earth-like planets in the galaxy from which inorganic raw materials could be obtained.

15.1.2 Theorem Between 5,748 and 9,642 animated Earth-like planets could exist in the galaxy from which organic raw materials could be obtained.

According to Theorem 2.4.3, the number of solar-like star systems with approximately Earth-great planets, in habitable zones, in our galaxy is **2.458 - 7.376 million**.
According to Theorem 2.5.3, the number of solar-like star systems in our galaxy, with approximately Earth-like planets in habitable zones, is **0.964 - 2.893 million**.
chapter 14 has shown that there are only a minimal number of civilizations. Thus, there are **several million** animated, approximately Earth-great and Earth-like planets from which raw materials could be obtained.

Then there are **billions** of planetoids, moons and asteroids that would be degradable.
When a species operates interstellar space flight, sourcing raw material from uninhabited planets or planetoids or moons should not be a major problem.
Whether it is about chemical elements, metals, ores, minerals, water, gases or even organic material such as wood, you do not need to approach (or conquer) inhabited planets.
There are enough uninhabited planets. And in some places even in higher concentration than on planets. There would be a reduction even more attractive. These include raw materials that would be interesting and accessible to us only in the future, such as helium 3 or methane.
Helium 3 could be extracted from lunar rocks in sufficient quantities. Methane and methane hydrate can be easily removed on Titan. And iridium, as well as the so-called rare Earths, are likely to be much richer on some asteroids than on Earth.

15.1.3 Theorem Raw materials are abundant in the galaxy.

15.2 - Maximum Distribution of Civilizations

The maximum number of civilizations in the galaxy results from the number of intelligent species. And that's when all the intelligent species would have made it to civilization.

Drake-corrected Special Basic Model
According to Theorem 9.7.2, the number of solar-like star systems in the galaxy, with an „Earth 2" in habitable zones inhabited by intelligent species, is **49 - 442** systems.

Seager-corrected Special Basic Model
According to Theorem 11.3.2, the number of solar-like star systems in the galaxy, with an „Earth 2" in habitable zone inhabited by intelligent species, is **77 - 742** systems.

15.2.1 Theorem   There could be a maximum of 49 to 742 civilizations on habitable „Earth 2", in solar-like star systems in the galaxy.

From the General Basic Model 8.4.2 the number N_{ieGal} of all star systems with habitable, Earth-like planets with intelligent species in the galaxy can be derived:

15.2.2 Equation $N_{ieGal} = A \cdot \Sigma(F_X \cdot F_{ph} \cdot F_{gae} \cdot F_{Li})$

$N_{ieGal} \approx A \cdot F_{ph} \cdot F_{gae} \cdot F_{Li}$

Insert the values from the corrected Basic Model into Equation 15.2.2:

$N_{ieGal} = (100 - 300) \cdot 10^9 \cdot 1:4{,}200 \cdot 1:214 \cdot 1:126$
$N_{ieGal} = 883 - 2{,}649$ intelligent species

$N_{ieGal} = (100 - 300) \cdot 10^9 \cdot 1:4{,}200 \cdot 1:1.078 \cdot 1:126$
$N_{ieGal} = 175 - 526$ intelligent species

The General Approach, equivalent to Equation 15.2.2, provides **189 - 2,664** intelligent species in the galaxy.

15.2.3 Theorem   There could be a maximum of 175 to 2,664 civilizations on habitable „Earth 2" in the galaxy.

15.3 - Distribution of Civilizations

The **Drake-corrected Special Basic Model** 9.8.1 provides **6 - 56** technological civilizations on Earth-like, habitable planets in solar-like star systems, in our galaxy.
The **Seager-corrected Special Basic Model** 11.4.1 provides **10 - 94** technological civilizations on Earth-like, habitable planets in solar-like star systems, in our galaxy.
The **transformed Seager-Equation** 10.3.2 provides **32 - 94** technological civilizations on habitable planets in solar-like star systems, in our galaxy.

15.3.1 Theorem   There could be 6 to 94 technological civilizations on habitable „Earth 2" in solar-like star systems in the galaxy.

The **corrected Drake-Equation** 9.5.3 provides **11 - 146** alien technological civilizations on habitable planets, in our galaxy.
The **Drake-corrected General Basic Model** 9.8.2 provides **22 - 199** technological civilizations on Earth-like, habitable planets in our galaxy.
The **Seager-corrected General Basic Model** 11.4.2 provides **35 - 334** technological civilizations on Earth-like, habitable planets in our galaxy.
The **General Approach** 12.2.2 provides **46 - 134** technological civilizations on Earth-like habitable planets, in our galaxy.
The **General Approach** 12.3.3 provides a maximum of **24 - 336** technological civilizations on Earth-like, habitable planets in our galaxy.

15.3.2 Theorem   There could be 11 to 336 technological civilizations on habitable „Earth 2" in the galaxy.

The **Drake-corrected General Basic Model** 9.9.1 provides **17 - 150** comparable civilizations on Earth-like, habitable planets in our galaxy.
The **Seager-corrected General Basic Model** 11.5.1 provides **26 - 252** comparable civilizations on Earth-like, habitable planets in our galaxy.
The **General Approach** 12.4.1 provides **18 - 254** comparable civilizations on Earth-like, habitable planets in our galaxy.

15.3.3 Theorem   There could be 17 to 254 comparable civilizations on habitable „Earth 2" in the galaxy.

The **Drake-corrected General Basic Model** 9.9.2 provides **6 - 49** space travelling civilizations on Earth-like, habitable planets in our galaxy.
The **Seager-corrected General Basic Model** 11.5.2 provides **9 - 82** space travelling civilizations on Earth-like, habitable planets in our galaxy.
The **General Approach** 12.4.2 provides **6 - 82** space travelling civilizations on Earth-like, habitable planets in our galaxy.

15.3.4 Theorem   There could be 6 to 82 space travelling civilizations on habitable „Earth 2" in the galaxy.

The **Drake-corrected General Basic Model** 9.9.3 provides **1 - 8** old civilizations on Earth-like, habitable planets in our galaxy.
The **Seager-corrected General Basic Model** 11.5.3 provides **2 - 14** old civilizations on Earth-like, habitable planets in our galaxy.
The **General Approach** 12.4.3 provides **1 - 14** old civilizations on Earth-like, habitable planets in our galaxy.

15.3.5 Theorem   There could be 1 to 14 old civilizations on habitable „Earth 2" in the galaxy.

The **corrected General Basic Model** 13.4.4 provides **3 - 48** humanoid civilizations on Earth-like, habitable planets, in our galaxy.
The **General Approach** 13.5.3 provides **4 - 48** humanoid civilizations on Earth-like, habitable planets, in our galaxy.

15.3.6 Theorem   In the galaxy, 3 to 48 humanoid civilizations could exist on habitable „Earth 2.

The **extended Seager-Equation** 10.2.1 for **Red Dwarfs** provides **14 - 40** technological civilizations in the galaxy.

According to Theorem 14.3.4, **approximately Earth-great planets** provides **47 - 141** technological civilizations in the galaxy.

15.4 - Result

As can be seen from the examples of the **M stars** and the **approximately Earth-great planets**, there are good chances for the existence of life, intelligence and civilization, despite small initial probabilities.
This is due to the large number of stars present. There are simply so many that even states of lower probability can manifest themselves in star systems and appear in sufficient numbers overall.

If life, intelligence and civilization in this universe have a probability **greater than zero**, then (due to the number of stars) there are also **several** technological civilizations - and not just one. The probability that a species is the only one in the universe is therefore close to zero, so it is **improbable**.

As can be seen in the previous treatise, although low but significant probabilities for the occurrence of life, intelligence and civilization can be deduced. So at least a few dozen technological civilizations are to be expected in our galaxy alone.
The basic data of this paper are based on the observations of the Kepler telescope, i.e. on the transit method. According to chapter 1.2, the geometric probability F_K of a transit (seen from the Earth) is 0.47 % = 1/213 %.
Therefore, there could be a maximum of about 213 times as many systems with planets as determined in the paper.
Extrapolated to the universe, it means that this universe is teeming with life, so to speak. life, intelligence and civilization are the **rule** and are accommodated as constant parts of the universe.

This results in an overall result:

> It is rather improbable that we are alone in the galaxy.
> It is more probable that we are not alone.

Then it can be concluded for the entire universe:

> It is improbable that we are alone in the universe.

15.5 - Building Set

Basically, the model presented so far consists of 6 equations.

The **Special Basic Model** for habitable „Earths 2" in solar-like star systems is:

6.3.3 Equation $N_{ze} = A \cdot F_{sph} \cdot F_{gae} \cdot F_{Liz}$

The **General Basic Model** for habitable „Earths 2" in the galaxy is:

8.4.2 Equation $N_{zexGal} = A \cdot \Sigma(F_X \cdot F_{ph} \cdot F_{gae} \cdot F_{Liz})$

The **General Approach** for habitable planets in the galaxy is:

12.2.2 Equation $N_{CivGal} = A \cdot \Sigma(F_X \cdot F_{ph} \cdot F_k \cdot F_{Liz})$

As well as the three equations, which contain all necessary probabilities concerning planetary adhesion, habitable zone, earth similarity, life, intelligence and civilization:

$F_{sph} = F_s \cdot F_p \cdot F_h$ Habitable Zone
$F_{gae} = F_g \cdot F_a \cdot F_e$ Earth Similarity
$F_{Liz} = F_L \cdot F_i \cdot F_z$ Civilization

All in all, the model presented here can be seen as a modular system. The model provides the basic building blocks and these can be adapted according to the problem.
Extensions can be used to identify old and humanoid civilizations. Adaptation can treat intelligent species or planets with life or planets about the size of the Earth in the galaxy. Also by adaptation the rate of civilization was determined, as well as by the cycle of civilization old civilizations could be defined.

Furthermore, the **Drake-Equation** and the **Seager-Equation** - after modification - are compatible to the presented **Basic Model.**

Thus, this building set provides a flexible repertoire of equations that are equivalent to each other in order to deal with the topic of civilizations in the galaxy in an exhaustive way.

	corrected special basic model	corrected general basic model	Drake-corrected gen. basic model	Seager-corrected gen. basic model	general approach
Earth 2	6.346 - 93.353	22.086 - 333.778	6.184 - 55.710	8.684 - 93.458	
Earth 2 with life	687 - 10.384	2.454 - 37.087	687 - 6.190	1.072 - 10.384	2.646 - 37.302
Earth 2 with intelligence	49 - 741	175 - 2.649	49 - 442	77 - 742	189 - 2.664
techn. civ. in sun-like systems	6 - 94		7 - 56	10 - 94	
technological civilizations		22 - 334	22 - 199	35 - 334	24 - 336
comparable civilizations	5 - 71	17 - 252	17 - 150	26 - 252	18 - 254
space travelling civilizations	2 - 23	6 - 82	6 - 49	9 - 82	6 - 82
old civilizations	1 - 4	1 - 14	1 - 8	2 - 14	1 - 14
humanoid civilizations	1 - 14	3 - 48	3 - 29	5 - 48	4 - 48

Appendix

16 – The SETI-Project

16.1 - The History of SETI

„*Search for Extraterrestrial Intelligence*" also called **SETI** describes the search for extraterrestrial civilizations. [1]
As early as 1909 Nikola Tesla was concerned with alleged signals from Mars. [2] In the same year, astronomer David Peck Todd unsuccessfully suggested searching for possible extraterrestrial radio signals using research balloons to carry receivers. [3] Guglielmo Marconi claimed to have received signals from aliens in the early 1920s, but this could not be confirmed. [4]

In September 1959, physicists Philip Morrison [5] and Giuseppe Cocconi [6] of Cornell University published a groundbreaking thesis in "Nature" entitled "Searching for Interstellar Communications". [7] They explain how radio astronomy could serve to receive potential interstellar communication. This publication is considered the birth of SETI.

At the same time as Morrison and Cocconi, but independently of them, the astronomer Frank Drake, [8] at the Green Bank radio Telescope in West Virginia, worked on an idea as to whether it might be feasible to receive signals from other worlds. The "Ozma Project" began in April 1960, initially with the observation of the two nearest stars to the sun, Tau Ceti and Epsilon Eridani.

In November 1960, scientists from various disciplines met for the first time in Green Bank, USA, to discuss the probability of extraterrestrial intelligences and the search for them. Participants of the conference included Frank Drake, Otto von Struve, Philip Morrison, Carl Sagan, Melvin Calvin, Bernard M. Oliver and John Lilly. [9]
During this conference Frank Drake also presented for the first time his calculation of the possible number of more highly developed civilizations in our galaxy, which has become famous in the meantime as **Drake-Equation**:

$$N = R \cdot f_p \cdot n \cdot f_L \cdot f_i \cdot f_c \cdot L$$

Nikolai Kardaschow, Josef Schklowski and other scientists organized further SETI conferences at the Byura-kan Observatory in 1964 and 1971. In 1966, Carl Sagan and Josef Schklowski published "Intelligent Life in the Universe", a widely known book on SETI. [10]

In 1971 NASA [11] also became interested in SETI with the "Project Cyclops". Plans for future research were developed during a summer workshop at Stanford University and NASA Ames Research Center.

From 1972 to 1976, under the name "Ozma II", the search for alien signals continued at the Green Bank radio Telescope. The researchers observed 674 stars during more than 500 hours of observation.

Carl Sagan, Bruce Murray and Louis Friedman founded the "Planetary Society" in 1980, which among other things was to financially support various SETI projects.

1984 the "SETI Institute" [12] was founded.

The declared goal was to promote and carry out research on SETI and life in the universe.
Among the founding members was Frank Drake and the radio astronaut Jill Tarter [13], who is regarded as a role model for the main actress in Carl Sagan's bestseller "Contact".

In October 1992, two parallel NASA-SETI programs began under the name "High Resolution Microwave Survey". (HRMS).

The astronomers of NASA Ames Research Center used the 305 - meter Radio Telescope in Arecibo for this purpose, targeting about a thousand previously selected stars.

While the researchers at the Jet Propulsion Laboratory (JPL) conducted a sky survey in the Mojave Desert using the 34 - meter Goldstone Telescope.
After less than a year of observations, $60 million in development costs and 23 years of planning, the NASA-SETI project was discontinued in 1993 following a savings decision by the US Congress. [14]

In February 1995, the SETI Institute launched the "Project Phoenix". The project was a further development of the NASA Ames Research Center's targeted search.

On the search list of the project were about 1,000, mostly son-like stars, in a maximum of 200 light-years away and a minimum age of three billion years.
Until September 1996, the 64 - meter Parkes - Radio Telescope served as a base in Australia.

In October 1995, with the support of the privately-founded Planetary Society, the Project BETA began work on the Harvard-Smithsonian Observatory's 28 - meter Telescope at Harvard, and in the counter theorem to Phoenix, BETA was not specifically targeted Investigate stars, but scour the entire sky on the frequency of neutral hydrogen.

In 1996 "Project Phönix" moved from Australia to Green Bank's 43 - meter Radio Telescope in Virgina.
At the same time, the receiver for the "SERENDIP" project (Search for Extraterrestrial Radio Emissions from Near-by Developed Intelligent Populations) was installed at Arecibo's 305 - meter Radio Telescope in Puerto Rico.
This SETI project, also supported by the Planetary Society, was similar to "BETA" based on the principle of non-targeted sky survey.

In 1998, "Project Phoenix" moved its main base to the 305 - meter Radio Telescope of Arecibo in Puerto Rico, where it is still stationed today.
At Harvard University and the University of California in Berkeley, optical SETI projects were launched for the first time. The Observatories searched specifically for very short flashes of light emanating from selected target stars.

The SETI@home project started in 1999. [15] It also used the SERENDIP receiver on the Arecibo Radio Telescope, but concentrated its search on a narrower frequency range. For the first time, the analysis of the data is not done on a stationary basis, but in the meantime on the more than **six million** desktop PCs of the participants, this worldwide Internet project.

In 2000, the optical SETI program at Harvard Observatory was expanded. It was to carry out the first complete optical sky survey by 2002.

Since 2001 Seth Shostak [16] Senior Astronomer at the SETI Institute. Headquartered in Mountain View, California, SETI Institute employs more than 50 researchers who explore all aspects of the search for life, its origin, the environment in which life develops, and its ultimate destiny.
Shostak is an active participant in the institute's observation programs and has been hosting SETI's weekly "Big Picture Science" radio program since 2002.

In March 2003, SETI@home astronomers used the Arecibo Radio Telescope for targeted, repeated observations of about 200 "candidate" signals. They were previously selected from the data of almost four years of work of the Internet project, as the most interesting and promising for a real extraterrestrial signal.

16.2 - Signals

To date, i.e. 2018, no relevant signal has been detected. Except for the so-called "Ohio-Wow-Signal". [17]

On 15 August 1977, the huge Big Ear Radio Telescope at Ohio State University registered the strongest and clearest potential alien signal in SETI's history. To this day, however, it has neither been observed again nor its cause clarified.

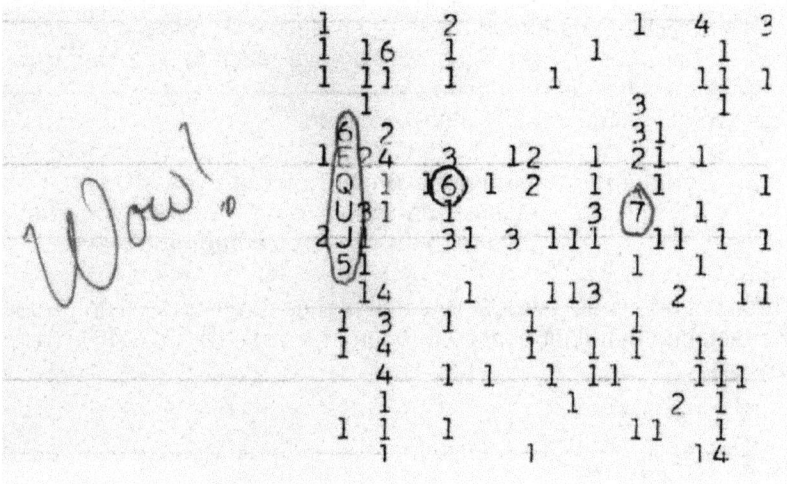

Sending signals to alien intelligence is called Active SETI or METI (Messaging to Extra-Terrestrial Intelligence) or CETI (Communication with extraterrestrial intelligence).

Researchers such as the astrophysicist Stephen Hawking or David Brin speculate, however, that Active SETI could be associated with considerable risks [18] [19] and that it would be appropriate to think about plans for planetary defense.

For the risk assessment of a transmitted signal, the ten-step San Marino scale was created, which ranges from insignificant (**1**) to extraordinary (**10**).

In 1974 a message was sent via the Arecibo antenna. The message was arranged in a matrix of 23×73 pixels containing information on numbers, chemical elements, nucleotides, DNA, humanity, planet Earth and the transmitter. According to the **San Marino scale**, the message had level 8 (far-reaching). [20]

16.3 - Operating Time of SETI

In 1895, Guglielmo Marconi [4] first used electromagnetic waves on our planet, about 120 years ago. 1901 saw the first transatlantic radio communication. From 1905 wireless radio traffic [21] was generally used, thus for 110 years.

In an orbit of 55 Lightyears, all these electromagnetic signals emitted by alien civilizations should have been detected to this day. And if they had answered, their signal would have reached us today. However, this has not yet happened.

The operating time of SETI is now 55 years.

16.3.1 Conclusion Within a radius of 55 Lightyears there is no civilization that can or wants to communicate with us via electromagnetic waves.

The Lightyear is an astronomical unit of measurement. A Lightyear is the distance that light passes through in a year. That's 9.461 trillion km. [22]

16.4 - No Answer

This can have several reasons:

1) There's no civilization within 55 Ly.
2) There is a civilization within a 55 Ly radius that is less developed than our civilization and does not yet know electromagnetic waves.
3) There is a civilization within a 55 Ly radius that is more advanced than our civilization and no longer uses electromagnetic waves.
4) There's a civilization within 55 Ly, but it won't communicate.

5) With SETI we hear in the wrong place, because **if a physics for interstellar space travel exists (Axiom 6.6.1), then there is also a message transmission on this basis, something like hyper radio. And this is certainly NOT electromagnetic and limited to the propagation speed of EM waves.**

SETI therefore behaves as if you are waiting for drum signs while the others are on the phone or in radio communication.
SETI has so far made it possible to study an area with a radius of 55 Lightyears. There are over 1,000 suns in an orbit of 55 Lightyears.
The only case left is where a civilization hundreds of years ago, a few hundred or a thousand Lightyears away, would have sent out a signal that would have reached us today. That would be an incredible stroke of luck.
We can rather assume that an interstellar message transmission, based on multidimensional physics, exists and is used by space faring species. We have to look for such a physics in order to communicate with another civilization in the galaxy.

16.5 - Quantum Technology

On 17 August 2016, the South German newspaper published an article about the launch of the Chinese satellite Mozi on 15 August 2016 from the Jiuquan spaceport, which is equipped with quantum technology for interception-secure communication - using entangled particles. Which, in the event of further development, enables tap-proof and instantaneous, i.e. over-light-fast, communication. [23] [24]
According to the South German newspaper, Austrian physicists succeeded in 2012 in transmitting such entangled light quanta over a distance of 143 kilometres between the Canary Islands of Tenerife and La Palma. This is the beginning of a tap-proof and ultra-fast communication, a technology that will probably be generally available within a few years. As expected, the world's military will be keenly interested in this technology.
It can therefore be assumed that such a type of communication is not only meaningful with interstellar journeys, but even standard. So when space travelling aliens use this technology for communication, a silence in space on the electromagnetic frequency band is understandable due to the eavesdropping security on the one hand and the super-rapid signal speed on the other.

This means that only civilizations of a certain stage of development use electromagnetic signals and thus betray themselves, so to speak.
With SETI it would therefore only be possible to receive signals from civilizations on the same or similar technological level as we are. Another indication of the "limitation" of the SETI approach. On the basis of the facts listed, the following can be formulated in general terms:

16.5.1 Theorem   **SETI, in its current form, is pointless because we are using the wrong methods.**

16.6 - Distribution of Starsystems

Our galaxy has a diameter a of about 100,000 Lightyears. The mean thickness **h**, the disk, is 3,000 Lightyears. This is also where most stars are. The center, for example in the form of a sphere, has a diameter **d** of 16,000 Lightyears. [20] The volume of the galaxy is thus approximated:

$$V = V_{sphere} + V_{cylinder} - V_{spherical\ segment}$$

$$V = \pi/6 \cdot d^3 + \pi \cdot a^2/4 \cdot h - \pi \cdot d^2/4 \cdot h$$

$$V = \pi/6 \cdot 16000^3 + \pi \cdot 100{,}000^2/4 \cdot 3000 - \pi \cdot 16{,}000^2/4 \cdot 3000$$
$$V = 2.510{,}341{,}97 \cdot 10^{13}\ [Ly^3]$$

If you take the space the galaxy occupies and transform it into a cube, the result is:

$$V = a^3$$

If this room is also occupied by a number **N** of stars, then this results:

$$V = a^3 \cdot N$$

Then results the **mean distance** between two stars to:

$$a = \sqrt[3]{\frac{V}{N}}$$

Our galaxy has about 100 to 300 billion suns. [25] There is an average distance of **4.37** to **6.3** Lightyears between the individual stars.

For comparison: Alpha Centauri, our nearest neighbour, is **4.24** Lightyears away.

This type of mean value determination, via the formation of cubes, is required for further considerations in order to be able to convert planetary distribution into mean distances. Due to the cube formation, the mean distances, between the stars, appear directly as edges of the cubes belonging to the suns.

The approach using a cube allows a relatively simple calculation of a "mean distance", which would be much more difficult with a cylinder or a sphere. When they are arranged one after the other, empty spaces are created that cannot be calculated, while cubes can be arranged seamlessly next to each other.

16.7 - The best Case

According to Theorem 3.3.1, **9,600 - 289,000** „Earths 2" exist in our galaxy.

At **9,600** „Earths 2", the mean distance is 1,378 Lightyears. That would also be the maximum mean distance to another civilization. With SETI, only a space of about 55 Lightyears was recorded. So we would have to send at least 1,323 more years or wait for the answer in 2,701 years.

At **289,000** „Earths 2", the minimum mean distance is 443 lightyears. In this case we would have to send or wait at least 388 years and in 831 years we would have the answer.

Overall, we would have to search on average between **831 to 2,701 years** with SETI and wait until we would receive an answer.

As a consequence, what tremendous luck one has to have in order to find extraterrestrial civilizations with the SETI Project.

According to Theorem 8.3.5, a maximum of **1 million** „Earths 2" could exist in our galaxy.

The mean distance between the planets is 293 Lightyears. In that case we would have to send at least 238 more years and wait and in **531 years** we would have the answer. As a consequence, signals from extraterrestrial intelligences are not expected to be recorded via the SETI Project in the next decades.

16.8 - Distances and Periods

The considerations so far refer to the best case, namely the maximum number of „Earths 2" in the galaxy. If you take the figures for civilizations, it looks even worse.

According to Theorem 15.2.3, a maximum number of **175 - 2,264** civilizations is to be expected.
With **175** civilizations, the mean distance is 5,235 Lightyears. That would also be the maximum mean distance to another civilization.
With SETI, only a space of about 55 Lightyears was recorded. So we would have to send or wait at least 5,180 more years, so we would get the answer in 10,415 years.
With 2,264 civilizations, the mean distance is 2,230 Lightyears. That would also be the maximum mean distance to another civilization.
With SETI, only a space of about 55 Lightyears was recorded. So we would have to send or wait at least 2,175 more years, so we would get the answer in 4,405 years.
Overall, we would have to search on average between **4,405** to **10,415** years with SETI and wait until we would receive an answer.

According to chapter 16.5, only civilizations of a certain developmental level use electromagnetic signals. According to Theorem 15.3.3, 17 - 254 comparable civilizations could exist on habitable „Earth 2" in the galaxy.
With **17** civilizations, the mean distance is 11,378 Lightyears. That would also be the maximum mean distance to another civilization.
With SETI, only a space of about 55 Lightyears was recorded. So we would have to send or wait at least 11,323 years to get the answer in 22,701 years.
With **254** civilizations, the mean distance is 4,623 Lightyears. That would also be the maximum mean distance to another civilization.
With SETI, only a space of about 55 Lightyears was recorded. So we would have to send or wait at least 4,568 more years, so we would get the answer in 9,191 years.
Overall, we would have to search on average between **9,191** to **22,701** years with SETI and wait until we would receive an answer.

16.9 - Consequences

A signal from another civilization would have taken a few thousand years according to chapters 16.7 and 16.8 and would also have had to be sent out at the „**right**" time to reach us today. It would be an incredible stroke of luck to receive such a signal with the SETI Project.

16.9.1 Theorem   **It is unlikely that SETI will receive signals from other civilizations in the coming decades.**

And if it were, as I have already said, an extraordinary stroke of luck for which there are only two possibilities:

1) There is a civilization only 50 to 55 Lightyears away.
2) It happens to be a signal that has been on the way for some time and was sent at the right time.

Another consequence results from the distances and/or the resulting periods of time. Communication with other civilizations is almost impossible. There's no point waiting a few centuries or millennia for an answer.

16.9.2 Theorem **Communication with other civilizations is almost unlikely due to the long transmission periods.**

So SETI can only serve to randomly receive a signal sometime, which can then confirm the existence of another civilization in the galaxy - and not more. Finding a civilization and referencing it within communication range would be an incredible stroke of luck.

According to Theorem 6.1.2, the 22nd century will be the time in which humanity will be able to conduct interstellar space travel and see for itself, so it is to be expected that SETI will be discontinued at the latest in the next century, or that **other methods of listening** will be available until then.

17 – The Fermi-Paradox

17.1 - The Considerations of Fermi

The Fermi paradox [1] was established in 1950 by the physicist Enrico Fermi [2]. He was concerned with the probability of intelligent alien life and therefore with the question: Are we humans the only technologically advanced civilization in the universe?
Due to the age of the universe and its high number of stars, intelligent life should also be possible and widespread outside the Earth.
Prerequisite is: The formation of life on Earth is not an unusual process or a galactic accident or isolated case. (see also axioms in chapter 4).

On the way to lunch at Los Alamos National Laboratory, in 1950, Enrico Fermi discussed this topic with Edward Teller, [3] Emil Konopinski [4] and Herbert York [5] on the basis of alleged UFO sightings. He asked himself: *"Why can neither spaceships of other space inhabitants nor other traces of extraterrestrial technologies be observed from Earth."* The paradox can therefore be represented as follows:
"The widespread belief that there are many technologically advanced civilizations in our universe, combined with our observations suggesting the opposite, is paradoxical and indicates that either our understanding or our observations are erroneous or incomplete."

Or in short: **If there are aliens, why haven't they already landed in public?**

17.2 - The Situation Today

This paradox was established in 1950. 65 years have now passed and the situation has changed decisively. Only a few UFO sightings were available in 1950. Today the number goes into the tens of thousands. The MUFON organization alone has documented over 70,000 cases. [6]
And part of it can be explained by the **extraterrestrial hypothesis**. [7]

According to Theorem 6.6.1, between **3 - 71** technological civilizations, on an "Earth 2", could exist in our galaxy, operating **interstellar space flight**.
According to Theorem 7.3.2, there could be between **1 - 12 old** technological civilizations in our galaxy that could have visited us. Besides, chapter 7 has shown that it is quite possible that we have already been visited by alien species in the past.
All in all, the observations mentioned in the paradox are incomplete.

And wouldn't it be rather naive to think that aliens present would also appear publicly? A similar situation arises here on Earth in primate research.
To study a horde of gorillas in their natural environment with their natural behaviour, one must remain invisible as an observer.
If you show yourself to the gorillas, the situation changes abruptly and the gorillas no longer behave naturally. The procedure of primate researchers consists of accompanying the monkeys and only when they have become accustomed to the visitor and return to their natural behaviour to begin the actual field research.
The situation is similar with the world's population and the aliens. If an alien species were to land here publicly, the entire psychological situation on Earth would be changed.
If aliens want to study us and conduct their experiments, it is appropriate to remain undiscovered.

A number of alien species are likely to be many times older than our civilization. In their eyes we would be better primates or just primitives.
Thus, the understanding mentioned in the paradox is incomplete or erroneous. Therefore, we can conclude here:

17.2.1 Theorem **The Fermi paradox is obsolete and can be omitted.**

The Fermi paradox is outdated in view of today's events and findings and the considerations in this book. (especially chapter 7.5).

Bibliography

1 – Planets in the Galaxy

1   https://en.wikipedia.org/wiki/Exoplanet

2   http://kepler.nasa.gov/

3   https://en.wikipedia.org/wiki/Kepler_(spacecraft)

4   NASA's Kepler Completes Prime Mission Begins Extended Mission. NASA 14. November 2012

5   https://en.wikipedia.org/wiki/Circumstellar_habitable_zone

6   Spiegel Online Wissenschaft Dienstag, 05.11.2013

7   NASA - Characteristics of Transits https://www.nasa.gov/kepler/overview/abouttransits

8   https://en.wikipedia.org/wiki/List_of_exoplanets

2 – Evaluation of Catalogue Data

1   http://phl.upr.edu/hec

2   http://phl.upr.edu/projects/habitable-exoplanets-catalog

3   http://www.astro.princeton.edu/~tdm/koi-fpp/ms.pdf

4   http://www.t-online.de/nachrichten/wissen/id_77804832/nasa-kepler-teleskop-entdeckt-fast-1300-neue-planeten.html

5   https://en.wikipedia.org/wiki/Exoplanet

4 – Animated Planets in the Galaxy

1   https://en.wikipedia.org/wiki/Circumstellar_habitable_zone

2   https://en.wikipedia.org/wiki/Earth´s_rotation

3   https://en.wikipedia.org/wiki/Earth´s_magnetic_field

4   http://www.magneticpulser.us/Publications_of_Dr_Wolfgang_Lud.html#SGROBJ7DB44EF22181671

5   Schumann, W.O.
    Über die strahlungslosen Eigenschwingungen einer leitenden Kugel, die von einer Luftschicht und einer Ionosphärenhülle umgeben ist
    Zeitschrift Naturforschung 7a, 149-154, 1954

6   Klaus Piontzik, Gitterstrukturen des Erdmagnetfeldes
    Books on Demand, Norderstedt S.57
    ISBN: 9-783833-491269

7   https://en.wikipedia.org/wiki/Atmosphere_of_Earth

8   Klaus Piontzik, Claude Bärtels, Planetare Systeme Bd. 1
    Books on Demand, Norderstedt S.116
    ISBN: 9783848232642

9   https://en.wikipedia.org/wiki/Greenhouse_effect

10 https://en.wikipedia.org/wiki/Ocean

11 https://en.wikipedia.org/wiki/Continent

12 https://en.wikipedia.org/wiki/Chemical_element

13 https://en.wikipedia.org/wiki/Amino_acid

5 – Intelligent Species in the Galaxy

1   https://en.wikipedia.org/wiki/Great_Oxygenation_Event

2   https://en.wikipedia.org/wiki/Symbiogenesis

3   https://en.wikipedia.org/wiki/Neoproterozoic

4   https://en.wikipedia.org/wiki/Snowball_Earth

5   https://en.wikipedia.org/wiki/Cambrian

6   https://en.wikipedia.org/wiki/Extinction_event

7   https://en.wikipedia.org/wiki/Ordovician

8	https://en.wikipedia.org/wiki/Devonian
9	https://en.wikipedia.org/wiki/Permian-Triassic_extinction_event
10	https://en.wikipedia.org/wiki/Siberian_Traps
11	https://en.wikipedia.org/wiki/Triassic
12	https://en.wikipedia.org/wiki/Cretaceous-Paleogene_extinction_event
13	https://en.wikipedia.org/wiki/Eocene
14	https://en.wikipedia.org/wiki/Oligocene
15	https://en.wikipedia.org/wiki/Priabonian
16	https://en.wikipedia.org/wiki/Rupelian
17	https://en.wikipedia.org/wiki/Eocene-Oligocene_extinction_event
18	https://en.wikipedia.org/wiki/Toba_catastrophe_theory
19	https://en.wikipedia.org/wiki/Pleistocene
20	https://en.wikipedia.org/wiki/Holocene
21	https://en.wikipedia.org/wiki/Cosmic_ray
22	https://en.wikipedia.org/wiki/Gamma-ray_burst
23	https://en.wikipedia.org/wiki/Supernova
24	https://en.wikipedia.org/wiki/Solar_flare
25	https://en.wikipedia.org/wiki/Asteroid
26	https://en.wikipedia.org/wiki/Comet
27	https://en.wikipedia.org/wiki/Planet#Planetary-mass_objects
28	https://en.wikipedia.org/wiki/Climate_change

29	https://en.wikipedia.org/wiki/Atmosphere_of_Earth
30	https://en.wikipedia.org/wiki/Sea_level
31	https://en.wikipedia.org/wiki/Volcanism
32	https://en.wikipedia.org/wiki/Supervolcano

6 – Civilizations in the Galaxy

1	https://en.wikipedia.org/wiki/Universe
2	https://en.wikipedia.org/wiki/Milky_Way
3	http://www.spiegel.de/wissenschaft/weltall/
4	https://en.wikipedia.org/wiki/Solar_System
5	https://en.wikipedia.org/wiki/Life
6	https://en.wikipedia.org/wiki/Earth
7	https://en.wikipedia.org/wiki/Homo_rudolfensis
8	https://en.wikipedia.org/wiki/Homo_habilis
9	https://en.wikipedia.org/wiki/Moon_landing
10	https://en.wikipedia.org/wiki/Exploration_of_North_America
11	https://en.wikipedia.org/wiki/Vasco_da_Gama
12	https://en.wikipedia.org/wiki/Ferdinand_Magellan
13	https://en.wikipedia.org/wiki/Francis_Drake
14	https://en.wikipedia.org/wiki/Martin_Luther
15	https://en.wikipedia.org/wiki/Galileo_Galilei
16	https://en.wikipedia.org/wiki/Johannes_Kepler
17	https://en.wikipedia.org/wiki/Telescope

18	https://en.wikipedia.org/wiki/Microscope
19	https://en.wikipedia.org/wiki/Renaissance
20	https://en.wikipedia.org/wiki/Middle_Ages
21	https://en.wikipedia.org/wiki/Bronze_Age
22	https://en.wikipedia.org/wiki/Iron_Age
23	https://en.wikipedia.org/wiki/Classical_antiquity
24	https://en.wikipedia.org/wiki/Agriculture
25	https://en.wikipedia.org/wiki/Ceramic_art
26	https://en.wikipedia.org/wiki/Cave_painting
27	https://en.wikipedia.org/wiki/Prehistory
28	https://en.wikipedia.org/wiki/Archaic_humans
29	Karl Popper. Logik der Forschung Akademie-Verlag, Berlin 2004, Herbert Keuth (Hrsg.)

7 – Survival of a Civilization

1	https://en.wikipedia.org/wiki/Overexploitation
2	https://en.wikipedia.org/wiki/Pollution
3	https://en.wikipedia.org/wiki/Human_overpopulation
4	https://en.wikipedia.org/wiki/Pandemic
5	https://en.wikipedia.org/wiki/War
6	https://en.wikipedia.org/wiki/Disaster
7	https://en.wikipedia.org/wiki/Revolution
8	https://en.wikipedia.org/wiki/Homo
9	https://de.wikipedia.org/wiki/Sternbildungsrate

10 https://en.wikipedia.org/wiki/Milky_Way

11 http://www.mufon.com/

8 – General Basic Model

1 https://en.wikipedia.org/wiki/Stellar_classification

2 https://en.wikipedia.org/wiki/Circumstellar_habitable_zone

3 G. Gonzalez u. a.: The Galactic Habitable Zone: Galactic Chemical Evolution. In: Icarus. Band 152, 2001, S. 185–200

4 N. Prantzos: On the "Galactic Habitable Zone". In: Space Science Reviews. Band 135, 2008, S. 313–322

5 https://en.wikipedia.org/wiki/Galactic_habitable_zone

9 – The Drake-Equation

1 https://en.wikipedia.org/wiki/Drake_equation

2 https://en.wikipedia.org/wiki/Frank_Drake

3 The Drake-Equation Revisited: Part I @wayback.archive.org. Astrobiology Magazine

4 https://de.wikipedia.org/wiki/Sternbildungsrate

5 Frank Drake, Dava Sobel, Signale von anderen Welten Droemer Knaur, 1998, ISBN 3426773511

6 https://en.wikipedia.org/wiki/Carl_Sagan

7 https://en.wikipedia.org/wiki/Search_for_extraterrestrial_intelligence

8 https://en.wikipedia.org/wiki/Voyager_1

9 https://en.wikipedia.org/wiki/Voyager_2

10 Carl Sagan
https://www.youtube.com/watch?v=MlikCebQSlY

10 – The Seager-Equation

1   https://en.wikipedia.org/wiki/Sara_Seager

2   https://en.wikipedia.org/wiki/James_Webb_Space_Telescope

3   https://en.wikipedia.org/wiki/Transiting_Exoplanet_Survey_Satellite

4   https://www.cfa.harvard.edu/events/2013/postkepler/Exoplanets_in_the_Post_Kepler_Era/Program_files/Seager.pdf

12 – A General Approach

1   https://en.wikipedia.org/wiki/Stellar_classification

2   https://en.wikipedia.org/wiki/Hertzsprung-Russell_diagram

13 – Lines of Evolution

1   https://en.wikipedia.org/wiki/Theropoda

2   https://en.wikipedia.org/wiki/Troodon

3   https://en.wikipedia.org/wiki/Dale_Russell

4   https://en.wikipedia.org/wiki/Human_evolution

5   https://en.wikipedia.org/wiki/Cambrian

6   https://en.wikipedia.org/wiki/Extinction_event

7   https://en.wikipedia.org/wiki/Devonian

8   https://en.wikipedia.org/wiki/Permian-Triassic_extinction_event

9   https://en.wikipedia.org/wiki/Triassic

10   https://en.wikipedia.org/wiki/Cretaceous-Paleogene_extinction_event

11   https://de.wikipedia.org/wiki/Konvergenztheorie_(Evolution)

12   https://en.wikipedia.org/wiki/Simon_Conway_Morris

16 – The SETI-Project

1   https://en.wikipedia.org/wiki/Search_for_extraterrestrial_intelligence

2   The New York Times, 23rd May, 1909 by Nikola Tesla, How to Signal to Mars

3   https://en.wikipedia.org/wiki/David_Peck_Todd

4   https://en.wikipedia.org/wiki/Guglielmo_Marconi

5   https://en.wikipedia.org/wiki/Philip_Morrison

6   https://en.wikipedia.org/wiki/Giuseppe_Cocconi

7   Searching for Interstellar Communications
Nature, Bd. 184, 1959, S. 844–846

8   https://en.wikipedia.org/wiki/Frank_Drake

9   Sebastian von Hoerner: Sind wir allein? – SETI und das Leben im All,
Beck, München 2003, ISBN 3-406-49431-5, S. 151–152

10   Iosif S. Šklovskij, Carl Sagan
Intelligent life in the Universe. Holden-Day
San Francisco 1966

11   https://www.nasa.gov/

12   https://en.wikipedia.org/wiki/SETI_Institute

13   https://en.wikipedia.org/wiki/Jill_Tarter

14   Searching for good Science: The Cancellation of NASA's SETI Program, Stephen J. Garber
Journal of the British Interplanetary Society,
Vol. 52, pp. 3-12, 1999

15   https://en.wikipedia.org/wiki/SETI@home

16	https://en.wikipedia.org/wiki/Seth_Shostak
17	https://en.wikipedia.org/wiki/Wow!_signal
18	Warnung von Astrophysiker Hawking Spiegel online, 25. April 2010
19	David Brin: The Dangers of First Contact, davidbrin.com
20	Iván Almár, Paul H. Shuch: The San Marino Scale: A new analytical tool for assessing transmission risk. Acta Astronautica, Vol.60, Issue 1, S. 57–59
21	https://en.wikipedia.org/wiki/Telegraphy
22	https://en.wikipedia.org/wiki/Light-year
23	http://www.sueddeutsche.de/wissen/satellitenstart-china-erprobt-quanten-kommunikation-im-weltall-1.3124422
24	http://www.spektrum.de/news/verschraenkte-photonen-aus-dem-all/1464637
25	https://en.wikipedia.org/wiki/Milky_Way

17 – The Fermi-Paradox

1	https://en.wikipedia.org/wiki/Fermi_paradox
2	https://en.wikipedia.org/wiki/Enrico_Fermi
3	https://en.wikipedia.org/wiki/Edward_Teller
4	https://en.wikipedia.org/wiki/Emil_Konopinski
5	https://en.wikipedia.org/wiki/Herbert_York
6	http://www.mufon.com/
7	Die Alien-Hypothese Klaus Piontzik, Claude Bärtels BoD Verlag

Images Directory

Page

10 https://de.wikipedia.org/wiki/Kepler_(Weltraumteleskop)

11,98 https://astrokramkiste.de/sonne

12,17 http://phl.upr.edu/projects/habitable-exoplanets-catalog
18,20,21,73,123

12 http://aasnova.org/2017/01/30/uv-habitable-zones-further-constrain-possible-life/

25,75 Recorded on the flight from Apollo 17 to the Moon on December 7, 1972

31 http://www.sternwarte-kraichtal.de/sonnensystem.html

32 http://ig-hutzi-spechtler.eu/aktuelles__planetentag.html

32 https://en.wikipedia.org/wiki/Moon

34 https://klexikon.zum.de/wiki/Atmosphäre

34 https://www.pinterest.de/pin/390968811372843403/

35 https://bildagentur.panthermedia.net/m/lizenzfreie-bilder/14125013/periodensystem-der-elemente-spanische-labeling/

35 http://www.seilnacht.com/Lexikon/amino.html

36 https://en.wikipedia.org/wiki/Primate

36 http://www.paguera.com/de/nachrichten/delfine-in-paguera-sehen

36 http://www.lingonetz.de/kids/wissen/Der-Rabe

36 https://www.helles-koepfchen.de/wissen/ratgeber-und-tipps/katze-als-haustier.html

Page

36	http://www.fotocommunity.de/photo/afrikanischer-elefant-ayubowan/35943842
36	https://www.bzfe.de/inhalt/schaf-und-ziegenrassen-vorgestellt-892.html
41	http://www.zum.de/Faecher/Materialien/beck/wplanet/wp/soncos.htm
41	https://www.forschung-und-wissen.de/nachrichten/astronomie/vor-1200-jahren-traf-ein-gammablitz-die-erde-13371816
41	http://www.supernovae.net/
42	http://www.vol.at/xxl-sonneneruption-magnetsturm-naehert-sich-der-erde/4082686
42	https://en.wikipedia.org/wiki/Asteroid
42	https://en.wikipedia.org/wiki/Comet
43	https://www.mpg.de/forschung/schneeball-erde-algen-eukaryot
43	https://www.stern.de/panorama/wissen/natur/themen/klimawandel-4170178.html
43	http://www.scinexx.de/dossier-701-1.html
43	http://energieinitiative.org/die-beweise-fuer-den-klimawandel/
44	http://www.spiegel.de/wissenschaft/natur/ecuador-vulkan-tungurahua-macht-feuerwerk-a-1080586.html
44	https://www.pravda-tv.com/2016/10/supervulkane-vorwarnzeit-hoechstens-ein-jahr-weltweite-vulkanaktivitaet-videos/
46	https://www.youtube.com/watch?v=BKDEnixO7po

Page

46,61 77,110	http://www.zeno.org/Meyers-1905/B/Dampfmaschine
46	https://www.flugrevue.de/raumfahrt/raumfahrt-auf-der-ila-2016/680326
62	https://commons.wikimedia.org/wiki/File:Lower_Manhattan _Skyline_from_Brooklyn_Heights_Promenade.jpg
64	http://www.sueddeutsche.de/wissen/bemannte-raumfahrt-umsonst-ins-all-1.1669125
73,107	https://en.wikipedia.org/wiki/Spiral_galaxy
83	https://en.wikipedia.org/wiki/Frank_Drake
86	https://en.wikipedia.org/wiki/Carl_Sagan
86	https://en.wikipedia.org/wiki/Voyager_1
95	http://www.skyandtelescope.com/astronomy-news/sara-seager-webinar-on-exoplanets/
95	https://en.wikipedia.org/wiki/James_Webb_Space_Telescope
95	https://www.wissenschaftaktll.de/artikel/NASA_gibt_ Startschuss_fuer_neuen_Planetenjaeger 1771015589052.html
107	https://de.wikipedia.org/wiki/Hertzsprung-Russell-Diagramm
116	https://www.schulentwicklung.nrw.de/sinus/front _content.php?idart=4349
117	http://menschnachmass.ch/
139	https://en.wikipedia.org/wiki/Green_Bank_Telescope
140	https://en.wikipedia.org/wiki/SETI_Institute

Page

140   https://www.mpifr-bonn.mpg.de/pressemeldungen/2014/8

141   https://de.wikipedia.org/wiki/Goldstone_Deep_Space_Communications_Complex

141   https://de.wikipedia.org/wiki/Parkes-Observatorium

141   https://www.cfa.harvard.edu/sao

142   https://setiathome.berkeley.edu/

143   http://www.bigear.org/

143   https://de.wikipedia.org/wiki/Wow!-Signal

150   https://www.nobelprize.org/nobel_prizes/physics/laureates/1938/fermi-bio.html

All other images are from the authors' archives

List of Names

Persons

	Page
Brin, David	143
Calvin, Melvin	139
Cocconi, Giuseppe	139
Da Gama, Vasco	52
Drake, Francis	52
Drake, Frank	82,139,140
Fermi, Enrico	150
Friedman, Louis	140
Galilei, Galileo	52
Hawking, Stephen	143
Janssen, Hans and Zacharias	52
Kardaschow, Nikolai	140
Kepler, Johannes	52
Kolumbus, Christoph	52
Konopinski, Emil	150
Lilly, John	139
Luther, Martin	52
Magellan, Fernando	52
Marconi, Guglielmo	139,144
Murray, Bruce	140
Morris, Simon Conway	117
Morrison, Philip	139
Oliver, Bernard M.	139
Petigura, Erik	10,11,14
Russell, Dale Alan	115
Sagan, Carl	86,87,139,140
Seager, Sara	95,96
Schklowski, Josef	140
Shostak, Seth	142
Struve, Otto von	139
Tarter, Jill	140
Teller, Edward	150
Tesla, Nikola	139
Todd, David Peck	139
Warburg, Otto	38
York, Herbert	150

Telescopes, Radio Telescopes | Page

Big Ear Radio Telescope	143
Goldstone Telescope	141
Green Bank Radio Telescope	139,140,141
Parkes-Radio-Telescope	141
Radio Telescope in Arecibo	140,141,142,144
SuperWASP	9
Very Large Telescope	10

Institutions

Cornell University	139
ESO	10
Habitable Exoplanets Catalog	15,17,18,20,21,123
Harvard-Smithsonian Observatory	140,142
Harvard University	142
Los Alamos National Laboratory	150
MUFON	72,150
NASA	10,15,140
NASA Ames Research Center	140,141
Ohio State University	143
Planetary Society	140,141
SETI@home	142
SETI Institut	140,142
Stanford University	140
University of California	10,142
Wikipedia	15

Satellites, Space Stations

ASTERIA	9
COROT	9
Hubble - Space Telescope	10,95
ISS	47,52,62
JWST	95,96,97
Kepler-Telescope	9,10,11,13,14,15,23 25,26,28,29,30,85 91,98,102,114
Mozi	145
TESS	95
Voyager 1	86
Voyager 2	86

Astronomy

	Page
Age of the Earth	46,150
Age of the Milky Way	46,70
Age of the Solar System	46
Age of the Universe	46
Alpha Centauri	147
approximately Earth-great planets	16,19,20,21,22,27 71,123
approximately Earth-like planets	16,19,20,21,22,25 26,70,131
Asteroids	34,40,42,52,115,131
Astrometric method	9
Biosignature	96,97,98
black holes	42
bolometric brightness	13
brown dwarfs	42
Chromosphere	42
Comets	34,40,42
Coronal mass ejection	42
Cosmic radiation	33,40,41,81
Direct observation	10
Earth 2	22,23,25
Earth axis	32
Earth's magnetic field	33
Epsilon Eridani	139
Exoplanet	9,10,15,16,30,95
Galactic habitable zone	80
Gamma flash	40,41
geometric probability	10
Ghost Planets	40,42
Gravity	18,19,23,123,129
G-Star	11,12,17,19,73,98 107,128
Habitable Planets	11,12,13,14
Habitable Zone	12,13
Hertzsprung-Russell-Diagram	107
Inclination	9,10
Kepler's orbit	42
Luminosity	13,41,107
Lunar landing	47,49,52,62
Magnetars	42
Magnetic field	33,34
Meteorites	39,40

	Page
Mars	24,34,52,139
mass	9,17,18,22,32,122
Milky Way	41,46,81,83
Moon	32,123,128,131
Moonstone	131
M-Star	96,97,98,107,108, 110,,120,128,135
Neutron stars	42
Number of stars in the Milky Way	12
Orbit	9,10,31,32,42
Orbit period	23,24
Oort Cloud	42
Planetary system	11,13,32,129
Planetoids	131
Prazession	32
Pulsars	42
Radial velocity method	9
Red Dwarfs	95,96,97,107,108 128,129,134
Rotation axis	32
Rotation period	21,23,24
Run-time of light method	10
sidereal month	32
Solar flare	40,42
Solar system	11,13,16,42,46,47 52,75,86,87
Spectral class	10,11,73,74,77,95 106,108,128
Star formation rate	70,81,83,90
Subearth	15,17,123
Sun-like	10,11,98
Superearth	15,18,123
Supernova	39,40,41,81
Tau Ceti	139
Temperature	10,13,16
Tide	32
Titan	131
Transit	9,10,98
Transit method	9
Universe	31,41,46,47,135,140 150
Venus	34

Epochs

	Page
Age of Enlightenment	47
Ancient Age	47,53
Baroque	47
Bronze Age	47,53
Cambrian	38,115
Chalk-tertiary boundary	39,116
Devon	39,115
Eocene	39
End Time Antiquity	47
Grande Coupure	39
Great oxygen disaster	38
Holocene	40
Ice Age	31,43
Iron Age	47,53
Kellwasser-Event	115
Middle Ages	47,52
Modern	47
modern times	47,52,53
Neoproterozoic	38
Oligocene	39
Ordovician	39
Permian triassic boundary	39,115
Pleistocene	40
Postmodernity	47
Priabonium	40
Renaissance	47,52
Rococo	47
Rupelium	40
Snowball Earth	38,40,43
Stone Age	47
Triass	39,116

Races

Anastasi	67
Babylon	47,53
Carthage	47,53
China	47,53,145
Egypt	47,53
Greek	47,53
Hittites	47,53
Indians	47,53,67

	Page
Israel	47,53
Kusch	47,53
Maya	67
Minoan	47,53
Romans	47,53
Sumer	47,53

Human Development

Agriculture	47,53
Ancient Races	47
Apes	46
Art	47
Australopithecus	46
Bone Flute	53
Cave Painting	53,72
Ceramic figures	53
Chimpanzees	46
Cities	47
Collector	47
Gibbons	46
Hominini	46
Homo erectus	47
Homo habilis	47,49
Homo rudolfensis	47,49
Homo sapiens	47,53,55,115
Humanoids	116,118
Human like	46
Hunter	47
Ivory carvings	53
Letterpress	47
Medicine	47
Meerkatzenverwandten	46
Megaliths	47
Neanderthal man	47,68
Primates	40,151
Science	47
Settlement	47,53
Stone Tools	47
Upright walk	46
Works of art	47

Locations	Page
Africa	67
America	40,52,67,72,86,95
Arecibo	140,141,142,144
Australia	40,141
Berkeley	10,142
California	10,142
Eurasia	40
Europe	10,52,53
Göbekli Tepe	47
Goldstone	141
Green Bank	83,139,140,141
Harvard	141,142
India	52
Indonesia	40
Jiuquan	145
La Palma	145
Los Alamos	150
Mojave Desert	141
Mountain View	142
Puerto Rico	141,142
Stonehenge	47
Sumatra	40
Tenerife	145
Toba Volcano	40
West Virginia	139
Wittenberg	52

Keyword Index

Keyword	Page
Age of a Civilization	68,89
Amino acids	31,35
Amphibian	116
Amphibians	39,115
Ancient Level	47
Argon	43
Atmosphere	18,23,32,33,34,38
Average development time	53,55
Average life span	54,68,89
Aviation	47
Basic Model, General	73,77
Basic Model, Special	60,66
Big Picture Science	42
Biochemistry	117,118
Biophysics	117
Birds	39
Brachiopods	38,39,115
Caldera	44
Carbon	31
Carbon dioxide	43
Chemicals	47
chemical elements	35,131,144
Chlorofluorocarbons	43
Circumnavigation	52
Civilization	46
Civilization cycle	71,89,136
Climate	31,32,34,35,38,39 40,43
Climate change	38,40,43,67
Comparable civilisations	62,79,94,105,111 34,149
Conodonts	38,115
Continents	23,34,35
Corals	115
Cosmic Level	47
Creodont	40
Cyanobacteria	38
Development level	46,47,48,49,52,56 58,60,61,63,68,125 126,146,148
Development level function	48,52,53,54,56,58
Development time	53

	Page
Dinosaurs	39,115,116
Disasters	38,40,56,67,115
Distribution model	30,49,66,121
DNA	35,38,117,144
Drake-Equation	82,139
Drake-Seager-Equation	95
Drinking water	35
Earthquake	67
Ebola	67
Electrical	47
Electric field	33
Electrons	41
Endosymbionic Theory	38
equipotential surface	43
Eukaryotes	38
Evolution	49,52,115,117,121
Exobiology	8,86
extraterrestrial civilizations	72,147
Extraterrestrial hypothesis	72,150
Fauna	23,35,40
Fermi paradox	150,151
Flora	18,19,23,35
Forest fires	67
Freshwater	34
Galactic level	47
Gases	131
General Approach	107
Genetics	118
genetic bottleneck	40
Geodynamo	34
Geomagnetic frequencies	33
Global warming	43
Gorilla	151
Green Bank Formula	83
Greenhouse effect	34,43
Hacker attacks	67
Helium 3	131
Hurricane	67
Hydrogen sulphide	39
Hyperfunk	145
Inorganic raw materials	130,131
Insects	39,115,119

	Page
Insectoids	116
Intelligent species	38,44,45,54,61,66 69,71,77,99,111,116 123,130,132,136
Internal conflicts	67
Interstellar Level	47
Interstellar space travel	47,49,64,65,72,85 131,145,149,150
Ionized atoms	41
Iridium	131
Lahar	67
Lens maker	52
level of origin	47
Life span of a civilization	53,54,55,68,84,89
Line of Evolution	116,118
Living things	33,34,35,38
mind-capable species	44,116
Magma chamber	44
Magma masses	33,35
Mammals	39,116
Mass extinction	39
Metals	47,131
Methane	43,131
Methane hydrate	131
METI	143
Middle Level	47
Minerals	35,131
Mitochondrion	38
Multiverse	47
Nitrogen	31,43
Nitrogen compounds	43
Nucleosynthesis	81
Nucleotides	144
Ocean	22,32,34
Old Civilizations	68
Ores	131
Organic material	131
Organic raw materials	130,131
overexploitation	67
Overpopulation	67
Oxygen	31,38,39,43
Ozone	43

	Page
Palaeotheries	40
Pangea	39
Pandemic	67
Periodic Table	35,116
Phosphates	31
Physics	47,64,145
Pisces	115
Plague	67
Plants	34,35,39,115
Planetary development dangers	40
Planetary level	47,49,52,72
Plate tectonics	33,35,67
Pollution	67
Prehistoric level	46
Pre Level	47,49,54,56
Primates	40,151
Project BETA	141
Project Cyclops	140
Project Ozma	139
Project Phönix	141
Prokaryotes	38
Protons	41
pyroclastic currents	67
Quantum Technology	145
Radio signals	139
Radio traffic	144,145
rare earths	131
Rate of civilization development	70,71
Raw materials	130,131
Reptiles	39,115
Reptilode	116
Revolution	68
Sabotage	67
Salts	35
San Marino scale	143,144
Sauroids	116
Schumann frequency	33
Sea level	38,41,43
SERENDIP receiver	141,142
SETI	63,86,139
SETI-Equation	83
Shipping	52

	Page
Siberian Trapp	39
Signal	10,33,63,139,140 142,143
Signal transmission	63
social unrest	68
Solar frequencies	33
Space Travelling	47
Space Travelling Civilization	64,79,80,94,105,111 134
Spanish flu	67
Statistics	30
Sulphur	31
Sulphur dioxide	39,43
Sum total	56,58,110
Supervolcano	41,44
Supply shortage	67
Technological civilization	61,77,84,96,98,109
Technological Level	47,72
Theory of Convergence	117
Therapsides	39,115
Theropods	115
Total development time	54
Total probability	58,60
Trace elements	31
Trilobites	38,115
Troodon	115
Tsuamis	67
Turn of the Era	52
UFO	72,150
UV radiation	34
Visitors	72,151
Volcanism	33,35,41,44,67
War	67 110
Water	13,31,32,34,39,131
Wether	31,32,34,35
Wood	131
Working hypothesis	121 40,41,81,96,98
Wow-Signal	143
Zika virus	67

www.ingramcontent.com/pod-product-compliance
Lightning Source LLC
Chambersburg PA
CBHW050215230526
45470CB00001B/391